No. 22.

DEPARTMENT OF DOCKS AND FERRIES
Pier "A," North River
NEW YORK CITY

REPORT

ON THE

Mechanical Equipment of New York Harbor

BY

B. F. CRESSON, JR., First Deputy Commissioner,

AND

CHARLES W. STANIFORD, Chief Engineer

SUBMITTED BY

CALVIN TOMKINS

Commissioner of Docks

DECEMBER 19, 1912

Report on the Mechanical Equipment of New York Harbor

January 24, 1913.

HON. WILLIAM J. GAYNOR,
Mayor of the City of New York.

Sir.—The increasing size of ships and the desirability of maintaining prompt sailings for liner service, which is supplanting tramp service at the important ports, has resulted in increasing necessity for prompt dispatch for terminals. Quick and cheap terminal service in loading and unloading vessels is essential to the prosperity of a port. The recognition for this need has suggested the submission of the accompanying report, prepared by First Deputy Commissioner, B. F. Cresson, Jr., and the Chief Engineer of the Department, Charles W. Staniford.

The Dock Department, in addition to controlling the municipal docks also exercises a large degree of control and supervision over private waterfront improvements, which can only be undertaken with the approval of this Department. The rapid development of public terminals more fully equipped than those which have hitherto been constructed, and the rapid development of privately owned waterfront makes discussion of port administration methods opportune.

The interdependence of port facilities makes it desirable that terminals shall be standardized as far as practicable to promote the coastwise trade of the country and to secure uniformity of terminal service at all our ports for the accommodation of ships engaged in oversea trade.

Respectfully submitted,

CALVIN TOMKINS,
Commissioner of Docks.

CITY OF NEW YORK.

DEPARTMENT OF DOCKS AND FERRIES.

NEW YORK, December 19, 1912.

REPORT ON THE MECHANICAL EQUIPMENT OF NEW YORK HARBOR.

CALVIN TOMKINS, ESQ.,
Commissioner of Docks.

SIR.—Many reports have been published by the different seaports, principally abroad, which have been profusely illustrated to show not only the characteristics of the harbors but also to bring into prominence the variety of mechanical equipment employed in the handling of freight. Very little has been written concerning the freight handling facilities in New York Harbor and it is the general impression among foreign engineers and others visiting this harbor, that New York is deficient both as to mechanical equipment as compared with foreign harbors and also as to the efficiency of the methods employed in handling cargo. There is probably no subject at the present time of greater importance in the science of harbor development, than a consideration of freight handling equipment, and it is the purpose of this report to describe the general equipment to be found about New York Harbor, and to discuss the methods of freight handling in order that it may form a basis for future consideration in the design of up-to-date terminals. Many photographs have been taken and certain drawings have been prepared for this report by the Staff of the Department to indicate the fact that in New York Harbor there is a great variety of freight handling devices, but it should also be borne in mind that in the handling of package freight on piers there has not been placed on any pier anywhere a mechanical equipment capable of taking the place of the cumbersome and expensive hand, horse or motor truck, and we believe that the consideration and development of some sufficiently elastic form of overhead carriage will be devised which will greatly decrease the cost of freight handling. One of the most important features to be considered in the design of modern dockage facilities is the rapid economical loading and discharging of cargoes carried by vessels entering the harbor. It is a question whether this important subject has been given the consideration it warrants in the many cases where terminal layouts have been planned and dock structures designed for the development of waterfront accommodations. Too often the pier or wharf under design has been considered mainly as a "structure along the waterfront to which vessels may be safely moored" and its cargo handling facilities have been considered of secondary importance, if considered at all. Vessels are brought to port to first embark or disembark passengers and second to discharge and load merchandise, either in bulk or packages, and all possible economy should be observed. The method of effecting these economies should be kept constantly in mind by the designer of modern pier or steamship terminals.

The cost to the shipper or manufacturer of handling merchandise by slow, cumbersome and expensive methods, is borne in fact by the consumer. Rapid mechanical handling will enable the manufacturer to sell cheaper and competition will quickly give to the consumer the benefit of lower cost to the manufacturer. If a truck with team and driver stands for half a day or more waiting for a chance to deposit its truck-load on a pier or on a railroad platform, the consumer pays for it in the higher price of the commodity. Delays to tugs, barges, lighters, derricks, etc., are all paid for by the consumer.

MECHANICAL EQUIPMENT IN USE IN NEW YORK HARBOR.

The criticism has been made that New York Harbor lacks modern methods of loading and discharging vessels. European engineers have pointed with pride to their great travelling cranes and derricks, rolling towers, etc., installed along the quays and pier sides at European ports, a notable case being the "Kuhwaerder" dock at Hamburg where one hundred and thirty-four great travelling cranes frown upon the harbor's shipping like the great guns of a fortress. European engineers have also criticized American methods of handling freight. To what extent this criticism is justified by the facts, is a question upon which shipping men differ and many authorities lean to the opinion that the American methods are no more expensive than those in vogue in Europe, when consideration is given to the cost, maintenance and depreciation of plant, difference in the rate of pay of labor, the number of working hours of the dock laborer and the difference in railroad rolling stock. Many authorities state that cargo is handled as rapidly in America, if not more so, than in Europe, particularly when sorting is required, which is nearly always the case with American freight, be it trans-Atlantic, Coastwise, Railroad or General.

For the handling of commodities in bulk, New York Harbor is probably better equipped than European ports. Great installations of the most modern and effective types of machinery for handling coal, grain, etc., in bulk will be found in the various railroad terminals, at the power plants of the electric lighting companies and manufacturing establishments. It is interesting to note further that the catalogues of American manufacturers of machinery show illustrations of many American hoists and conveying machinery which have been installed by them in European countries.

European ports are frequently of such dimensions that the eye of the observer can readily mass the entire harbor with its cranes, derricks and vessels into a single picture, presenting to the eye a whole which lends itself readily to a visual analysis of its various parts. This is not the case with New York Harbor, where five great cities—Manhattan, Brooklyn, Queens, the Bronx, Richmond, forming Greater New York, and the New Jersey District, are combined into a great metropolis, with a shore front of 748 miles.

An exact analysis of the amount of waterfront, together with the amount of developed wharfage in the whole Port of New York, as well as the developed wharfage in the City of New York proper, may be of interest and is tabulated, as follows:

New York Wharfage Table.

The straight waterfront in Greater New York is......555 miles.
The straight waterfront in New Jersey is............193 miles.

Total waterfront (straight) in Port of New York is 748 miles.

In Greater New York	224 miles	of wharfage has been developed of a straight waterfront of 100 miles.
In New Jersey	133 miles	of wharfage has been developed out of a straight waterfront of 20 miles.
Total,	357 miles	of wharfage has been developed in the Port of New York, out of a straight waterfront of 120 miles.

The City of New York has developed 53 miles of wharfage room, including 232 City owned piers.

The prudent observer would hesitate to form a quick conclusion regarding the equipment distributed over this expanse of shore front for use commercially or industrially. The combined crane and derrick installations of the principal ports of Europe could be distributed over this immense waterfront without materially changing its skyline or general physical aspect to the eye of the traveller sailing up the Bay in the modern thousand foot passenger steamship. The combined navies of the world would find safe anchorage in deep water within the limits of this harbor. At a recent naval demonstration or rendezvous the ships of the United States Navy, one hundred and thirty-one in number, of all classes from the largest warship to the smallest torpedo boat were gathered in the Hudson River in the upper Manhattan waters. Their position had to be made known to the average New Yorker by the daily newspapers and the above mentioned traveller would not see them until his vessel had proceeded ten miles up the Bay and North River. A trip along the immense waterfront of the North River, East River, Long Island Sound, the Staten Island and New Jersey shore, will show the observer that many types of freight handling machinery or apparatus used in the old world are in use in New York Harbor. There is, however, a notable difference in location or the placing of these mechanical devices. Where cranes, derricks, elevators, coal pockets either movable or stationary, are found, they are not installed at great expense in what may be termed batteries extending along the entire sides of piers, or along extended portions of the quay or bulkhead sea wall, but they are placed at such points along the waterfront or adjacent thereto where their greater reach and capacity are required for a definite function, as for example—handling heavy bulk at railroad terminals, power plants, manufactories, coal plants, machine shops, etc. It is impossible to give a complete description of each of the numerous power installations for handling cargoes within the limits of a necessarily brief report of this kind. There are given below, however, numerous photographic illustrations of the more important structures and apparatus used. Hundreds of minor structures such as railroad gravity coal pockets, industrial lifting and hoisting apparatus of smaller establishments, have not been shown, owing to the lack of space.

EUROPEAN METHODS.

An examination of the methods in use in handling ship's cargo in Europe develops the fact that notwithstanding the extensive and expensive installations of crane and derrick batteries at most of these ports, the following conditions generally obtain:

1. The most general use of the crane or derrick is to lift cargo from the hold of the ship to the deck of the pier or quay, from which point it is moved by hand truck, horse dray, etc., to the warehouse, railroad car, or consignee's space; a small proportion only being handled directly from ship to railroad car.

2. The ship's winches are very generally used notwithstanding the installation of cranes and derricks, and particularly are these used in handling cargo over the side to barges, which is extensively done at such ports as Rotterdam and Hamburg where a great deal of freight is transshipped by barges into the interior of the country on the Rhine and Elbe Rivers. At Antwerp one of the steamship lines uses the ship's winches exclusively notwithstanding the fact

that an extensive crane equipment has been supplied by the Municipality.

3. Few mechanical devices are found at European ports to move freight over the surfaces of the pier deck after the crane has deposited it. Two and four wheeled hand trucks and horse drays are used almost entirely for this movement and motor trucks have not been used to any extent. Portable electric winches also are scarcely used at all. It should be remembered in this connection that the great crane and derrick installations at European ports are generally made by and at the expense of the Municipal or State Governments having jurisdiction or control over the harbor, and not at the expense of the steamship companies occupying the piers. These derricks and cranes are rented out to the steamship companies with an operator, very much the same as the pier sheds or wharf space is rented out.

It would seem from the foregoing that as far as the freight handling facilities at steamship piers are concerned New York Harbor differs from European ports only in that it has no great crane or derrick batteries and further that in the matter of machinery, appliances or methods to reduce the expenses incident to sorting and distributing cargo over the pier deck or quay area, which is the greatest cost factor in the price per ton of handling freight, European ports are at least as lacking in them as is New York.

CLASSIFICATION OF MARINE SHIPPING.

The shipping of New York Harbor may be divided into six classes:

1. The trans-Atlantic passenger line service carrying package freight.
2. Trans-Atlantic and general sea-going steamship service not dependent on passenger schedule—this includes tramps.
3. West Indian and South American service.
4. Coastwise service.

5. River, harbor and Sound lines for passengers and freight.
6. The general harbor transportation business including all types of lighters, barges, canal-boats, railroad car floats, floating grain elevators, coal loaders, etc.

A correct conception as to what facilities would be necessary to more economically handle these various classes of freight must be based upon an intimate knowledge of the natural movement of freight in reaching and leaving the harbor waterfront, methods of conveying to and from the ship, railroad warehouse, etc., final conclusions and recommendations being based upon an analysis of the existing conditions and present methods of freight handling in New York Harbor.

CARGO HOISTS AND METHOD OF OPERATION IN NEW YORK HARBOR.

Practically all merchandise or freight whether in our out bound, reaching the steamship pier in New York Harbor must be sorted for various purposes. Incoming freight must be sorted by the marks for weighing, gauging, customs inspection, delivery to consignee, etc. Outgoing freight must be sorted and stored on the dock according to its bulk, perishable nature, etc. Where a vessel makes several ports it is necessary also to sort the cargo by ports. A large amount of bulk cargo for export does not come on the pier at all but is loaded directly into the ship from a lighter. The conditions governing the movement of package freight are in the main so nearly identical for all classes of vessels that the methods of handling employed differ only in the minor modifications of the same system. Practically identical methods are employed by all the trans-Atlantic lines in loading and discharging cargo and in coaling. Some of the coastwise, South American and West Indian lines have special features not applicable to the larger vessels and these will be referred to later. The piers in New York Harbor in many instances are supplied with steam or electric portable or stationary winches, placed at convenient intervals along the sides of the pier for use opposite the ship's hatches. Cargo masts or cargo hoists are placed along the sides of the piers, and the following is a brief comparison of the methods employed in European ports and in New York Harbor.

In most of the European ports, the great bulk of freight which is being moved from the ships hold to the quay or pier, or the reverse movement, is done by means of dock cranes. These cranes are operated either by steam, electricity or hydraulically, and they are usually able to move a certain distance along the pier in order to get the most advantageous location for the hatch to which they are working. Frequently the ship's winches are used to hoist the cargo from the hold to the deck, and from this point the dock crane picks the cargo up and places it on the pier. In many instances, however, the dock crane reaches directly into the hold, and in one movement lifts the cargo, swings it and places it on the dock. This involves the swinging of the boom of the crane, and there is a certain loss of time which must be experienced in locating the boom in the correct position to lift the cargo and then swing it into the correct position in order to deposit it.

The general practice in New York Harbor, in the newer pier installations, is to place what is known as cargo hoists along the sides of the piers. These are nothing more than structural steel bridges extending above the sides of a pier, to which may be attached pulleys or blocks practically at any point along the side of the pier (see Fig. 1).

The method of handling cargo is to rig one of these blocks directly over the door on the pier through which freight is being handled, and to place a block on one of the ship's masts directly over the hatch in the ship which is being worked. A line is passed from a winch, either on the ship's deck or on the pier, through a block directly over the hatch, and this line is attached beyond the block to another line which is run from a winch through a block on the cargo hoist of the pier, and near the point of junction of these lines there is placed a hook. In discharging cargo from the hold of a ship, the line passing through the block on the pier is left free and the cargo is lifted out of the hold by the winch operating through the block on the ship's mast above the hatch. When the cargo is clear of the ship's hatch and deck, the line from the winch operating through the block of the cargo hoist on the pier is brought into action by slackening the line passing through the ship's mast block and by taking up on the line passing through the cargo hoist block, the cargo is brought directly beneath the block on the cargo mast and is lowered from that to the deck of the pier. This is an exceedingly simple and effective method of handling cargo. It is very positive in its action, for when one line only is carrying

FIG. 1. Cunard Line S.S. "Carpathia" at Pier 54, North River, loading cargo by combined ship and dock winches.

the hook is directly in position over the hatch and when the other line is carrying, the hook is directly over the door in the pier shed. There is no swinging of booms to be carefully gauged, and there is no doubt of the location where the cargo will be when either one line or the other is carrying.

In short, nothing moves but the line and the great advantage of this has been recognized, not only by the steamship companies, but also by those who have carefully observed the loading and discharging of cargoes by the dock crane method and by

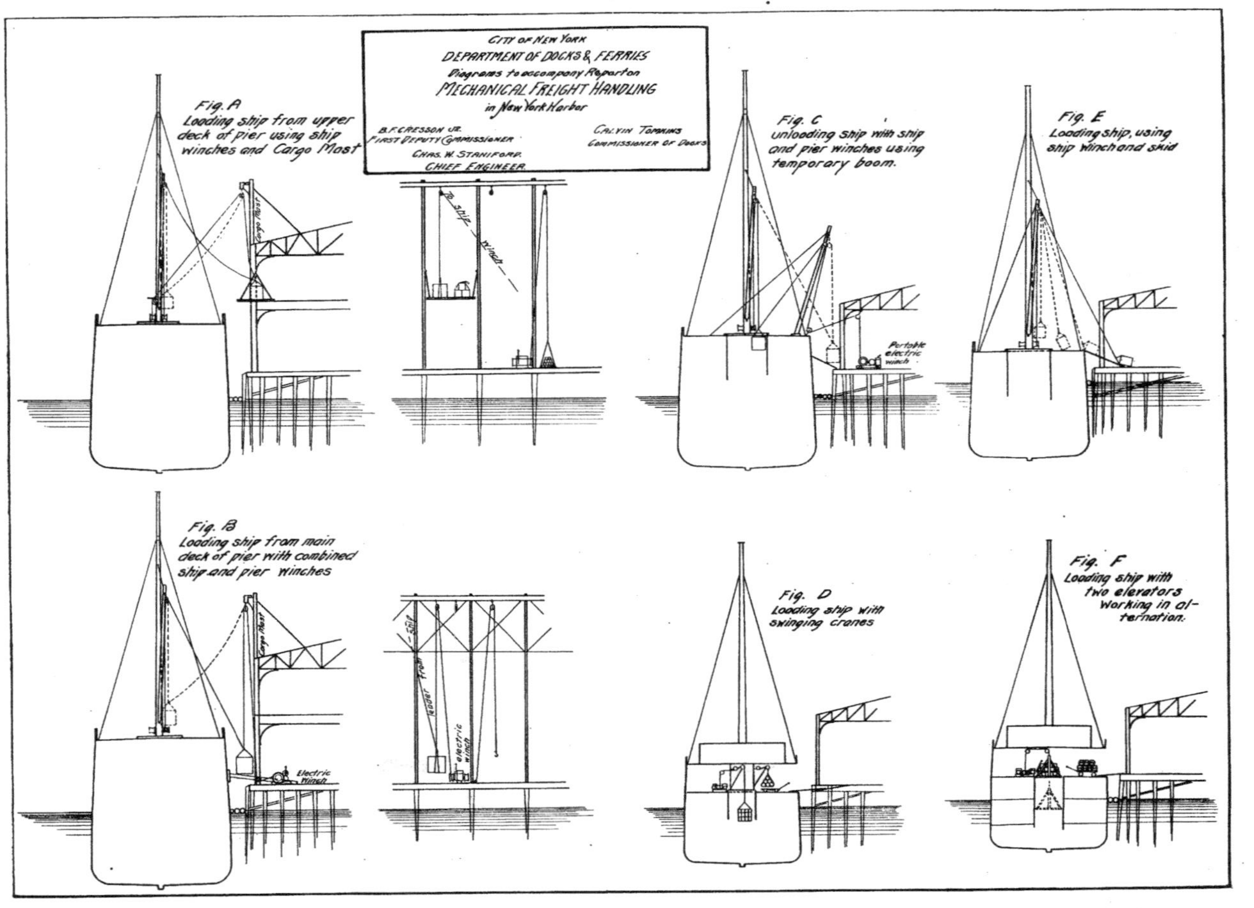

CITY OF NEW YORK
DEPARTMENT OF DOCKS & FERRIES
Diagrams to accompany Report on
MECHANICAL FREIGHT HANDLING
in New York Harbor
B. F. CRESSON JR.
FIRST DEPUTY COMMISSIONER
CHAS. W. STANIFORD
CHIEF ENGINEER
CALVIN TOMKINS
COMMISSIONER OF DOCKS
Fig. A
Loading ship from upper deck of pier using ship winches and Cargo Mast
Cargo Mast
To ship winch
Fig. C
unloading ship with ship and pier winches using temporary boom.
Portable electric winch
Fig. E
Loading ship, using ship winch and skid
Fig. B
Loading ship from main deck of pier with combined ship and pier winches
Cargo Mast
Electric Winch
Ship
from
leader
electric winch
Fig. D
Loading ship with swinging cranes
Fig. F
Loading ship with two elevators working in alternation.

Fig. 2. Chelsea Piers, North River, showing dock winch and ship's winch hoisting freight.

Fig. 3.

Fig. 4. Chelsea Piers, North River. Discharging cargo by combined use of ship's and dock winches.

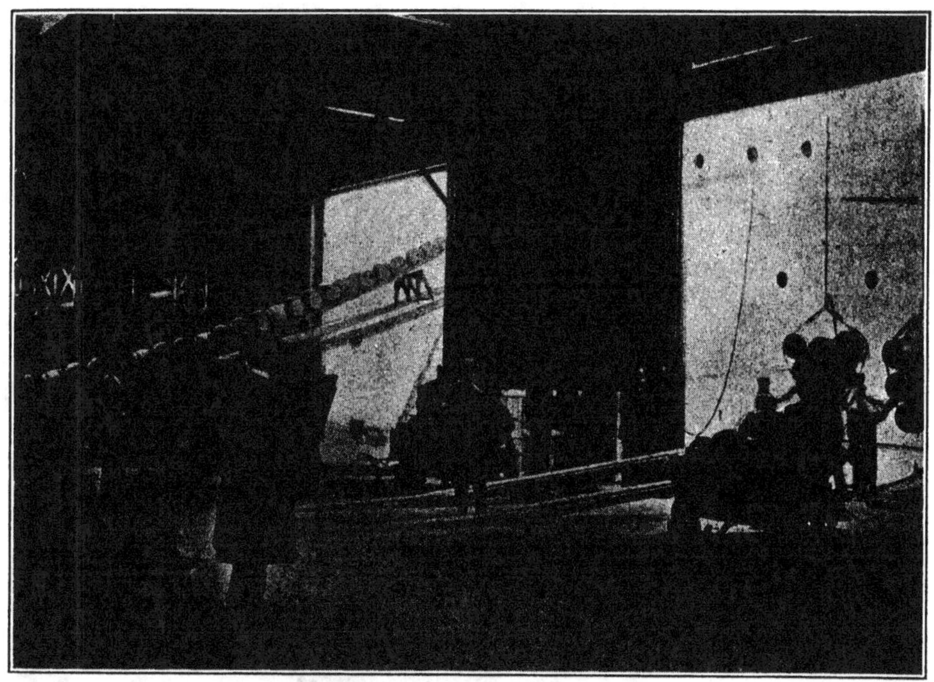

FIG. 5. Portable Electric Dock Winches, 20 H.P., unloading cargo. Fabre Line, foot 31st Street, Brooklyn.

FIG. 6. Electric Dock Winch. Holland-American Line. 18 H.P. Direct-current motor.

FIG. 7. Coaling Ship. Portable bucket (conveyors) elevator on endless chain suspended from side of pier by boom. Swing chute is attached to elevator, discharging directly into bunkers. Elevator is driven by small electric motor connected to upper driving wheel. Current is brought from the pier by detachable wire, which can be "plugged" in at switches in place along the dock shed. This method in general use on passenger lines. (See diagram.)

FIG. 8. Coaling Ship. Portable Conveyors. (See Fig. 7.)

FIG. 9. Coaling Ship. Floating Coal Elevator "Pardee." (See Fig. 10.)

FIG. 10. Floating Coal Elevator "Pardee." Penna., Beach Creek & Eastern Coal Co. There are five of these elevators in use in the harbor. Capacity, 1000 tons; handles 100 tons per hour.

FIG. 11. Coaling Ship. By use of electric dock winch and temporary mast rigged up on upper deck of ship.

FIG. 12. Coaling Ship. Barges loaded with coal brought alongside. Floating steam hoister lifts dump buckets rigged to temporary boom on ship's upper deck.

the use of ship's mast and cargo hoist. Foreign engineers and others who have at first commented upon the inadequate methods of handling cargo in New York Harbor, when our methods have been drawn to their attention have usually expressed surprise, and later great interest in the excellence of the methods employed here. It is significant also that in constructing piers in this country by some of the foreign steamship lines, they have equipped these piers with cargo hoists and portable winches in preference to installing dock cranes, such as have been provided for them by the port authorities at their terminals in Europe.

On very many piers in New York Harbor, however, the cargo hoists have not been placed, particularly in the older piers; and sometimes masts have been erected along the sides of piers with cable between, extending along the side of the pier, to which to attach blocks; where neither of these have been provided, it is the custom for one of the ship's masts to be rigged directly over the door of the pier so that the same method of operation is carried on by means of the ship's tackle itself. It is not unusual to use one of the ship's masts to discharge cargo by lifting from the hatch with one of the ship's masts and swing it onto the pier by the motion of the mast. Various modifications of this general method, which is believed to be the best one in use, and these, in a general way are shown in the diagrams on page 12. It is believed, that although the cargo hoist method may not be able to lift quite as large a load as the dock crane, yet by its use cargo can be discharged or loaded with greater rapidity than by any other present known method.

Most of the dock winches are double drum with "nigger heads" at each end, permitting the operator to lift two loads with one dock winch. Their size varies from 18 to 35 horse power. The capacity varies from 20 to 40 cycles per hour with loads ranging from 300 to 2,000 pounds.

While this loading is proceeding on the dock side of the ship frequently coaling is done from barges lying on the water side of the ship or between the ship and the side of the pier. While cargo is being discharged from the ship onto the pier or the reverse movement, floating grain elevators, derrick lighters and barges are loading to or from the ship on the water side (see Figs. 13 and 14). Coaling may also be under way and this is generally done by floating coal elevators and

FIG. 13. Type of Floating Derrick in general use in New York Harbor for loading and discharging cargoes in bulk.

FIG 14. Ship loading from water-side by floating steam derrick.

Fig. 14A. Floating Grain Elevator

bucket conveyors, which mechanically lift coal from the barge and deposit it directly into the ship with little or no manual labor. Incoming freight is wheeled generally by hand truck to space reserved on the pier deck for the various marks or consignees, for customs inspection, weighing, gauging, or railroad shipments, etc. Some steamship lines occupying two or more adjacent piers deposit incoming freight on the deck of one pier, and the vessel is then warped across the slip to the second pier upon which outbound freight has been accumulated. Outgoing freight reaches the pier by horse dray, railroad car float, lighter or barge with or without derrick, steam floating derrick, etc., and a considerable amount of sorting is done on the pier. Reference has been made to sorting for different ports and in several cases notably the Tyser Line to Australia, touching at ten ports, an elaborate system of marking cargo has been initiated, the pier being divided into ten or more zones for receipt of cargo for the various ports to be touched. Since comparatively few railroad connections to and upon the piers and bulkheads exist, the reach inboard of a derrick boom would be of value only in depositing freight brought from the hold, over a greater area of the deck surface, or picking up outgoing freight from a greater area. The quantity placed in the hold, however, is limited by the space available to deposit it before stowing, and the number of men who can work to advantage in the hold without interference. The same applies to unloading. The gang on the pier can generally work faster than that in the hold of the ship. The loading and discharging of vessels in New York Harbor is generally done by contract with a stevedore at a fixed price per ton. This price varies from 28 to 40 cents a ton and a good general average approximates 35 cents per ton, and by far the greater portion of the cost is occasioned by the necessary sorting of the freight by expensive hand trucking over the deck of the pier.

COASTWISE, ETC.

The American-owned ships engaged in the coastwise, Cuban, West Indian and South American service are different in some respects from the trans-Atlantic vessels. Many of these vessels have from three to four decks with large side ports leading to

FIG. 15. Typical Coastwise Steamship. S.S. "Saratoga." N. Y. & Cuba Mail S.S. Co. Note high side rail opened, allowing freight to be hauled to hatch over skid by ship's winch.

FIG. 16. Old Dominion Line. Loading steamer through lower ports.

the second deck, which permits direct trucking by hand or auto trucks from the deck of the vessel to the pier, the side port being not at a greatly different level from the deck of the pier (see Figs. 15 and 16).

Elevators and small power cranes lift freight to the various deck levels and are often supplemented by additional winches on the main deck with derrick booms rigged to the ship's masts for direct loading to and from floating lighters or other vessels. All of this apparatus may be used simultaneously. The Old Dominion Line occupying piers 25 and 26 North River, plying between New York and Norfolk, is particularly well equipped. Storage battery trucks run from the pier deck into the steamers through the side ports and electric dock winches are used also in loading and unloading. The piers are two-story structures, well equipped with elevators, reversible escalators with power cranes on the upper decks to load trucks standing on the lower deck, chutes from the upper deck to load drays standing on the bulkhead, etc. Electric dock winches are used to load and unload lighters and scows which are not equipped with power. There are daily sailings which necessitates the loading and discharging of about 3300 tons of merchandise in about fifteen hours at an average cost of about 35 cents per ton. About one million tons of freight is handled here annually.

River, Sound and Harbor freight reaching the piers is handled almost entirely by hand truck and dray. Much of this freight is, however, handled from and to lighters and steam derricks alongside the vessels. Many of these vessels load on the main deck only, this deck being enclosed by housing, large doors or ports permitting trucking into what might be termed the "hold" thus formed.

RAILROAD FREIGHT.

In the case of railroad freight, as with almost all other freight handled in the Harbor, much sorting is necessary. A railroad freight station generally consists of one or more piers adjacent to or connected by a bulkhead shed with receiving platforms.

Railroad cars with freight for Manhattan are placed on car floats at the New Jersey terminals and floated over to the rail-

Fig. 17. Receiving Station for West Bound Freight. D., L. & W. R.R., North River.

Fig. 18. Canal Street Freight Station, D., L. & W. R.R. All freight is hand-trucked through open door, shown on right, to cars standing on car-transfer float.

road piers in Manhattan, usually during the night or early morning. The freight is usually taken out of the cars by hand trucks and stacked on the piers and much hand-trucking is here necessary. The consignee is allowed a certain length of time, free of storage, in order to permit him to come for his freight, after which time the railroads usually take the freight to some ware-

Fig. 19. Loading Merchandise Into Railroad Cars Standing on Car-Transfer Floats.

house and charge the consignee with the expense. Much delay and expense results to the consignee owing to the crowded conditions of the piers and to the congestion not only of freight but of trucks. West Bound freight is usually delivered to the bulkhead or receiving platform, where the freight is weighed, sorted and marked for a certain car. The car floats are backed up against the bulkhead, sometimes two or three end on, and freight is wheeled from the bulkhead shed by hand truck to the particular cars which are marked for some city or distributing station along the line of the railroad. It is frequently necessary to wheel by hand truck freight as much as 700 feet along the central gangways of these car floats. It would seem that the installation of mechanical handling apparatus at these stations together with the rearrangement of methods of receipt and delivery by cooperation between the shipper and the railroad company would result not only in direct saving and expense to the companies, but to the manufacturer as well, by avoiding the unnecessarily long waiting of trucks and teams along the marginal way, due to their inability to discharge their loads.

Careful consideration should be given in this connection to the fact of the needless exposure for hours of drivers and animals to the severe winter weather prevailing along the waterfront of New York Harbor, which probably annually kills many men and horses engaged in the trucking business. Many are driven to the saloons to keep warmth in their bodies, and in this condition they are unable to give full value of their services to their employers. As a result, more teams and more men must be employed to perform the same amount of labor and in this way the consumer pays the higher cost of the commodity. This is a matter for serious consideration and cannot be regarded lightly.

FREIGHT HANDLING AT MARINE TERMINALS.

Many private terminal companies are operating throughout the Harbor and these companies handle freight with every one of the New Jersey Railroads as well as with all of the steamship lines, and in the case of railroad freight they usually have transfer bridges in order to bring the railroad cars on land, where the freight may be unloaded and where the cars may again be re-loaded. Notable among these companies are the Bush Terminal Company, The New York Dock Company, The Jay Street Terminal, The Brooklyn Eastern District Company and several smaller companies in the Harlem District.

In the Borough of Brooklyn, two great steamship terminals have been organized through private enterprise—The New York Dock Company and the Bush Terminal Company.

The New York Dock Company's plant extends in practically

an unbroken stretch along the East River and Buttermilk Channel from the Brooklyn Bridge on the north to the Erie Basin on the south, a distance of about three miles. Forty piers and a large grain elevator are built along this waterfront paralleled by two hundred warehouse buildings for storage, a large part of which is devoted to the storage of Brazilian coffee for which this terminal is considered the market. New modern reinforced concrete buildings are now in course of construction for warehousing and factory purposes and cold storage is being provided. A double track marginal railroad with car float bridges for direct connection with the trunk lines in New Jersey and Manhattan is laid along the marginal street extending along most of this development between the piers and the warehouse buildings, but this railroad system is divided into three parts; by Atlantic Avenue and Hamilton, over which no direct rail connection is made.

In the improvements contemplated by this Department, elevated connection is provided for over these avenues, so that there will be an unbroken line of rails over the whole frontage.

The following are the principal steamship lines located at this terminal:

Booth S. S. Line, to Brazilian Ports.
Lamport & Holt Line, to Brazilian Ports and Argentina.
Red "D" Line, to Venezuela, Porto Rico.
N. Y. & Cuba Mail S. S. Co., to Cuba and Mexico.
Pierce Line, to Mediterranean Ports.
Anchor Line, to Mediterranean Ports.
Trinidad Shipping & Trading Co., to West India.
Insular Line, to Porto Rico.
Clyde Line, to West Indies.
N. Y. & Porto Rico S. S. Co.
Uranium S. S. Co., to Hamburg.
N. Y. & Demarara S. S. Co., to West Indies.
Prince Line, to Brazil.
Funch Edye & Co., to Brazil.
Tyser Line, to Australian Ports.

The principal cargoes consist of coffee and sugar in bags, cocoa and Brazil nuts, and the Mediterranean vessels carry large quantities of macaroni, fruits, wines, etc.

Southerly of Gowanus Canal are located the two largest piers in the Harbor, one at the foot of 31st Street—1475 feet in length, leased to the Fabre Line operating to Mediterranean Ports and the other at the foot of 33d Street—1616 feet in length, owned by the City of New York for general wharfage. These piers are owned and were built by the City and are equipped with single story sheds and in the case of the latter, with a sprinkler system.

The Bush Terminal Company's plant lies south of these piers and the piers extend from the foot of 40th Street to 51st Street. Buildings for warehousing and manufacturing, which are a part of this great terminal, are located inshore and extend, or are planned to extend, from 28th Street to 51st Street. This is the most complete terminal development in the Harbor and probably on this continent. The seven piers average 1250 feet in length. A marginal railroad has been constructed connecting on the waterfront with transfer bridges having a classification and assembly yard, and with spur connections to the piers, to the warehouses and to the factory buildings in the rear, tying the whole together as a unit. Modern reinforced concrete factory buildings with power have been constructed in this terminal and there are at present about one hundred and fifty-five different kinds of manufactories in these buildings. The piers are occupied by six regular lines to Europe and South American ports and many so-called tramp steamers are berthed here.

Steamship lines docking here are as follows:

Austro-American S. S. Co., N. Y. & Mediterranean Ports.
Norton & Son, N. Y. & South American Ports.
Funch Edye & Co., N. Y. & West Indian Ports.
Prince Line, N. Y. & Calcutta, Bombay, etc.
Lloyd Brazileiro, N. Y. & Brazil.
American-Hawaiian Line, N. Y. & San Francisco & Hawaii by transshipment through the Isthmus of Tehuantepec.

The principal cargoes consist of coffee, dried fruits, cocoa and citric fruits, etc., from the Mediterranean.

The general method of loading and discharging steamers at the terminals of the New York Dock Company and of the Bush Terminal Company is practically the same as that described before. Portable electric dock winches are used to assist the ship's winches and to unload lighters and barges. At the Bush

Terminal the incoming freight is received in the following average proportions:

Destined for the warehouse, about....................	30%
Destined for railroad shipment in carload lots, about....	25%
Destined for railroad shipment in less than carload lots, about	25%
Destined for dray removal to consignee, about..........	20%
Total.....................	100%

It will be seen, therefore, that only about 25% of the incoming freight could be landed directly from the ship to the railroad car and taking into account the delay caused by the customs inspections, sorting, etc., it is an open question whether special machinery could be worked to advantage, and while the derrick or crane would be useful for this service, they could not be used elsewhere. Freight destined for warehouses is hauled usually on low flat horse trucks (see Fig. 20).

FIG. 20. Flat Horse Trucks. Bush Terminal and N. Y. Dock Co., Brooklyn.

One horse will keep three of these trucks continuously under way; two horses will move five trucks. While the horse or mule is under way with the second truck, the first is being unloaded by the warehouse "whip" and the third is being loaded at the barge, lighter, railroad car or ship and the operation is very nearly continuous. While this method appears to be cumbersome and obsolete, it is more economical here than the modern

FIG. 21. Storage Battery Crane Truck with Trailers. Bush Terminal Co., Brooklyn. Lift capacity of crane truck, 2,000 lbs.

electric crane truck with trailers. One of these electric motor crane trucks at a cost of $2500 plus the salary of the driver, plus depreciation, etc., would be required to perform the same amount of work as the horse and truck, aggregating $500 in cost. The crane trucks with trailers are here used mostly to load or unload cars where heavy lifting is required or on long hauls between dock and warehouse.

The warehouse buildings are five or six stories in height and are equipped with freight elevators and with bag and barrel chutes. The use of elevators for freight has been practically discontinued, for the reason that it is slow and uneconomical. Most of the freight is hoisted to the warehouse floors by the use of "whips," that having been found the most economical and rapid method. Electric winches are placed on the marginal road in front of the center of each of the warehouses and about 75 feet

FIG. 22. Showing "Whip" Hoists, operated by electric dock winches. Bush Terminal Co., Brooklyn.

from the building and each winch controls from two to four "whips" (see Figs. 22 and 23).

Eight hundred bags of coffee averaging 130 pounds each have been raised by a single "whip" in an hour.

Investigations made by the Bush Terminal Company during a number of years with a view of cheapening the cost by the installation of mechanical appliances and machinery, have not resulted in the placing of any machinery whatever, the conclusion being, apparently, that on account of the great amount of sorting required no present known form of installation would have sufficient elasticity to be economical. The electric motor crane truck with trailers, however, has been found generally efficient, especially for lifting heavy weights and loading cars for long hauls between dock and warehouse.

The New York Dock Company has installed at the foot of

FIG. 23. Showing Portable Electric Winches Operating "Whips." Bush Terminal Co., Brooklyn.

Baltic Street a telpher system for transporting freight between a railroad station, a warehouse and lighters. The installation is not extensive enough to demonstrate the efficiency or economy of more extensive terminal equipments of this type.

INSPECTION OF THE PLANT MENTIONED AND DISCUSSIONS ON THE SPOT with officials in charge have brought forth the following statements and data:

The Bush Terminal Company states that an investigation covering a period of several years for the purpose of installing cargo handling machinery has resulted in the conclusion, that present mechanical devices lack flexibility and would not be economical to install on account of the great amount of sorting required. The hand truck is the best for short hauls. Storage battery crane

FIG. 24. Telfer for Freight Transfer from Lighters to Warehouse and Railroad Cars. Note raisable boom at bulkhead end of bridge.

trucks of 2000 pounds capacity with trailers are economical for long hauls, for loading cars, etc., and where lifting is required. The "whip" operated from an electric dock winch is the most rapid and economical method to convey freight from the dock level to the warehouse floor. The horse or mule serving three trucks at one time is more economical than the auto-truck for average length of haul. No derricks or cranes are required; they would be in the way. The hand-truck is considered the best method of handling in sorting freight on the pier for average distances.

The New York Dock Company is not extending its small telpher and transpher installation. Lack of flexibility is stated to be the cause. The new buildings under construction are not to be provided with mechanical devices other than elevators and electric winches for "whips." Derricks or Gantry cranes, travelling along the piers, would have no function other than the ship's winch. Electric motor trucks are good for long hauls or bulk; horse and hand-trucks are generally cheaper on short hauls and on account of the great amount of sorting of the various marks required.

The Old Dominion S. S. Co., has installed a telpherage system at their Richmond Terminal; this has been found of use only in handling a single commodity when shipped in bulk in similar packages. It is not economical to handle general package freight by it. At their New York piers, electric storage battery trucks of 2000 pounds capacity handle about one-eighth of the 1,000,000 tons per annum handled by the company; they are used on long hauls only and for large consignments of similar packages of small bulk. The piers are two-story, equipped with adjustable ramps, many electric winches, elevators, reversible escalators, chutes to loading drays—thereby utilizing the second story to fullest capacity. It is intended in the future to equip the ramps with the "Reno" type of hand-truck conveyor, consisting of projecting hooks travelling on an endless chain, engaging the hand-truck axle and pulling it up the ramp incline. The most efficient method of moving over the surface of the pier under the present conditions of freight receipt is the hand-truck. Derricks or cranes are not wanted; their function is the same as the dock winch. The Company has decided that telpher or transfer are not flexible enough to be economical on account of the great amount of sorting required. Reversible escalators, running by steam power, handle one hundred and fifty bales of cotton per hour against fifty bales by elevator method formerly used to lift to second floor. Lighters are loaded and discharged by aid of electric dock winches installed on raised platforms.

Ships have side ports for loading to lower decks, enabling auto-trucks to travel directly into the hold to and from the dock.

Two electric auto-trucks of 2000 tons capacity are in use

Fig. 25. Elevated Electric Hoisting Winches. Old Dominion S. S. Co., Pier 26, N. R. These winches are placed on raised platforms near door-openings in the pier-shed. Rope hoists lead to blocks suspended from roof trusses. Used to load and discharge cargo and for loading and discharging lighters and barges without power. Capacity, 2½ tons. 10-H. P. motors.

Transfer Pier 6 Erie R.R. Jersey City, N.J.

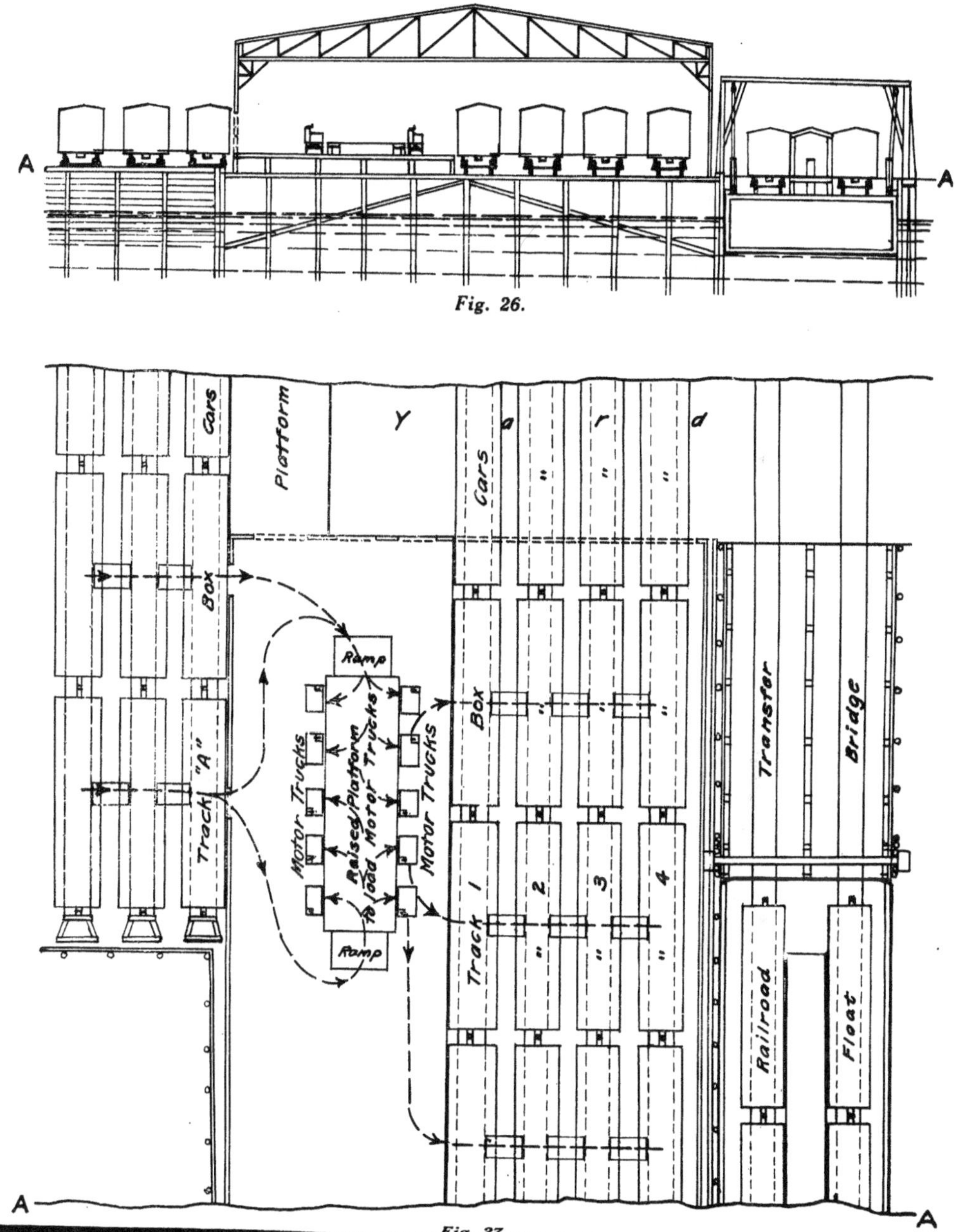

Fig. 26.

Fig. 27.

by the Mallory and Clyde Lines (coastwise) for handling bulk or large consignments of similar packages on long hauls only, as, for example, oil in barrels from the ship's hold direct to the marginal roadway, a haul of six to eight hundred feet. A small proportion of freight only is handled by the trucks.

Pier 20, North River (Erie Railroad), is used exclusively to handle fruit; the great bulk of California fruit is discharged here; this arrives generally in carload lots and is unloaded from the car, standing on the car transfer float, to the dock by hand truck and sorted by consignment. Gravity roller conveyors were used experimentally and discarded as not economical. Auto-trucks were tried experimentally and discarded for the main reason that the platform extending longitudinally along the center of the car floats were too narrow to allow two auto-trucks to pass.

Pier 6, Jersey City (Erie Railroad) is used as a transfer station for mainly West-bound freight, and the system of operation should be carefully noted. The operations here will no doubt furnish important facts which will be useful in the final solution of the mechanical distribution on the pier deck. Excellent results have been obtained here by the introduction of electric storage battery trucks of about 2000 pounds capacity. The use of twenty of these trucks has resulted in a reduction of labor required from eighty to about thirty-five men.

Figs. 26 and 27 show a layout of the pier. Freight is brought to the pier on track No. 1, carted by hand-trucks to the sorting platform, on both sides of which are assembled the electric trucks labelled with a number corresponding to a similar label hung up on the various cars or groups of cars, standing on four tracks on the opposite side of the pier. Communication over the four tracks is maintained by placing box car doors opposite, thus forming a transverse passage. The auto-truck goes inside of the car to discharge its load. The approximate reduction in cost of handling is about 9 cents per ton; about 700 tons of freight are handled daily. It would seem that an extension of this plan could be adapted to general railroad freight handling in New York City in connection with car float delivery.

Pier F, Erie R. R., Weehawken, N. J., is used for general package freight. The pier is two-story, equipped with barrel elevators and steam platform elevators. The roller conveyor is used to some extent for unloading packages from cars. Will not handle heavy bulk; as a rule hand trucking is more economical than the roller conveyor.

Erie R. R., Pier H, Weehawken. This is an open pier for general freight. Tracks run the full length of the pier. General merchandise is loaded to and from cars and lighters, barges, etc., by 15-ton locomotive cranes running along the length of the pier. A twenty-ton locomotive crane loads and discharges cars and barges at the bulkhead on the south side of the pier.

D., L. & W. R. R., Pier 9, Hoboken. This is a reinforced-concrete two-story building. Barrel elevators, helical and straight incline bag and barrel chutes and hydraulic elevators are used here. Three 2000-lb. capacity storage battery trucks load and discharge directly from cars. The pier is used for flour handling.

Corn Products Co., Shady Side, opp. 100th St., N. Y. The factories of the company are located approximately 1000 feet from the waterfront. Practically all the products of the establishment, syrup in cases, barrels, sugar in bags, starch, etc., are carried to the river front by a moving floor conveyor or "endless freight carrier." The carrier is 2200 feet in length and consists essentially of a moving floor extending from the factory to the waterfront; travelling at the rate of about 50 feet per minute. Boxes, barrels, bags, etc., are simply placed on this floor, no further handling is required until the waterfront is reached. The entire apparatus is driven by a 30 H.P. Electric Motor.

Midland Linseed Oil Co., Shady Side, opp. 100th St. This company is erecting a new steel hoisting tower at the outer end of its pier with an endless freight carrier over 2000 feet long leading from the tower to the factory buildings 1000 feet or more inshore.

GENERAL.

As has been stated before, the principal cost of handling freight is the trucking of it about the pier. The necessary rehandling incident to this is a heavy item of expense also. The great difficulty about mechanical appliances in their present form is their lack of elasticity of movement and until a perfectly elastic system and one which is not unduly complicated is found, little can be done to lessen the cost of present methods. Motor trucks and travelling platforms have been applied and have been successful to some extent, but they do not wholly solve the problem of economical freight handling about the pier or between the

FIG. 28. Electric Storage Battery Trucks. Erie R.R. Transfer, Pier 6, Jersey City. Twenty of these trucks are in use there. Capacity, 2,000 lbs.

pier and the back lands. Telpherage or overhead conveyance has been tried on a few piers in this country but so far has not proven successful, in its present form.

This may be due to improper design or it may be that the principle is unsuited to the character of business done on the pier. It is recognized that it is very difficult to design any form of overhead conveyance which will be sufficiently elastic to meet the requirements of general commerce, but if a form of overhead conveyance can be designed which will be thoroughly elastic as well as not overcomplicated—that will receive freight from any point on a pier and transport it in one movement to any other point on that pier or to any building adjacent on the upland and to any floor in that building, it would seem that this would go a long way toward cutting out the hand-trucking which, as before stated, is the principal item of expense. The possibility is recognized of great economies being effected along this line and mechanical engineers and operating men generally are giving attention to the problem of lessening these handling costs.

GENERAL CONCLUSIONS AND REMARKS.

1. It can be safely stated from the foregoing that the great movable derrick or crane installed along the docks and quays in Europe are not adapted to economical handling of American package freight under the present system of pier construction in New York Harbor.

2. It can be shown that the greatest cost factor in freight handling at steamship terminals is the trucking required to sort the package freight, after it has reached the pier or quay level. Europe is as bad off, if not worse off, than the United States, in mechanical means of performing this part of the work.

3. It is in this requirement, to distribute over the pier deck, that the greatest saving in cost can be made. Contemplated new pier construction should be planned to facilitate this movement over the pier deck, means being provided to utilize both decks of two-story piers to the fullest advantage.

4. It is important to remember that the modern tendency is to consider the water front of a harbor a connected whole, a terminal or group of terminals, planned for the benefit of the community at large. This may result in the connection of piers or groups of piers by marginal railroad tracks, making loading to and from cars a more common method of handling cargo than it is at the present time, and resulting in a perhaps economical installation of Gantry cranes and derricks.

5. A reluctance to depart from old and tried methods, together with the increased cost, seems to be a stumbling block to the introduction of improvements in many instances. New de-

vices are received with suspicion, particularly by employees, and are often not given a fair trial. Lack of study of the necessities of the particular class of business results often in poorly planned installations.

6. The modern telpher and transfer may in many instances be made adaptable to the movement over the surface of the quay or pier deck. If combined with reversible escalators and chutes, an efficient economical method of handling may also, in some instances, be evolved by a study of the requirements of the particular steamship or railroad line involved. By adjustment between the shipper and carrier of methods of delivery and removal of goods from the pier, such installation could in many instances be much simplified.

7. Assuming that the length and physical construction of a pier is fixed by the length of the vessel it is to berth, the next most important feature to be considered in the design or layout is the economical and rapid handling of cargo reaching it.

THE PRIMARY PURPOSE of this report being to show what methods are in use in New York Harbor in freight handling, no recommendations are herein made for more extensive use of any particular type or class of the many excellent mechanical devices and apparatus manufactured. The foregoing general description of the functions required point the way, however, to the intelligent study of the requirements of each particular pier or group of piers.

Efficiency data obtained by close observation and classification of the movement of freight at a particular locality, where mechanical installation is contemplated, the tabulation of these movements under the classification of unavoidable, avoidable, unnecessary, obstructive, etc., with the careful determination of cost of each movement per ton handled, should readily show whether the possible saving in cost per ton, considered as interest on the proposed investment, would be sufficient to warrant mechanical installation.

There can be no question that the time has come for a more extended use of economical means of cargo handling. The loss in money, due to delays, caused by the congestion on piers, bulkheads, etc., with present methods, is enormous. A reduction in cost of handling after reaching the pier of only two cents per ton would result in an annual saving of half a million dollars on the foreign commerce alone.

ILLUSTRATIONS.

The City of New York, through its Department of Docks, exercises general control of all construction work undertaken along the waterfront of New York Harbor whether it be undertaken by private parties or public authorities. Plans for all construction work of whatever nature contemplated by private or public interest along the waterfront are first submitted to the Department of Docks and Ferries for approval. These are carefully examined for structural stability, location with respect to established lines of improvement, public health, etc., etc., by the engineers of the Department, permit for erection of the work being granted if found to conform to established rules. Many hundreds of plans of structures are annually examined and reported upon.

The following pages illustrate some of the types of fixed portable and floating mechanical apparatus, derricks, elevators, etc., found. Some of these have been further illustrated by diagrammatic sketches showing the scheme of operation as far as possible. It will be noted that cranes, of one or more types are in use by all the railroads in the harbor for handling general freight. Most of the more prominent installations are for handling materials in bulk, generally coal. Following these illustrations is given a list of the various types of harbor craft in use for freight transportation.

Acknowledgment is made to Philip Guise, Assistant Engineer, for valuable assistance in the preparation of this report.

Respectfully submitted,

B. F. Cresson, Jr.,
1st Deputy Commissioner.

Chas. W. Staniford,
Chief Engineer.

December 23, 1912.

TYPES OF MECHANICAL EQUIPMENT IN USE IN NEW YORK HARBOR

Fig. 29. Diag. 29A. Travelling Cranes, Penna. R.R. Co., Greenville Terminal. Handles general freight from and to cars and vessels. Each crane spans three railroad tracks, and moves along dock under its own power. Capacity, 10 tons, by electric motors; booms can be raised in vertical plane.

Fig. 30. Diag. 30A. Travelling Direct Unloader. West Shore R.R., Weehawken. Movable booms, independent hoist, trolley, A. C. electric motors. Handles loads 5 to 10 tons on each boom.

Fig. 31. 20-ton Gantry Crane, D., L. & W. R.R., Hoboken.
Diag. 34A.

Fig. 32. 30-ton Locomotive Crane Used for Freight Handling. Erie R.R., Weehawken.

Fig. 33. 15-ton Locomotive Crane for Freight Handling. Erie R.R., Weehawken.

Fig. 34. Diag. 31A. Stationary Crane with Trolley Fall. Penna. R.R. yard, Hoboken. Crane is stationary, blocks suspended from trolley run between two box girders. Electric power.

FIG. 35. DIAG. 35A. Locomotive Crane, Pier No. 4, D., L. & W. R.R., Hoboken. Loads and unloads package freight directly from and into cars and vessels. Three of the cranes are at present in use. Capacity is 15 to 20 tons.

FIG. 36. Locomotive Crane, Brooklyn Rapid Transit Co., 65th Street, Brooklyn, Terminal. Electric; capacity, 15 tons; 4 of these cranes in use.

Fig. 37. Method of Handling Paper Pulp by Locomotive Crane. Tidewater Paper
Diag. 37A. Mills Co., foot of 30th Street, South Brooklyn.

Fig. 38. Travelling Coal Hoister and Cable Railway, Orford Copper Co., Constable Hook. Steam tower with grab bucket, double-cylinder steam engines, 10" x 24". Makes 3 cycles per minute. Handles about 1,000 tons per day. Length of tramway, 2,500 feet; **double track.**

FIG. 39. Railroad Car-float Transfer Bridge. Electric operation. B. & O. R.R., St. George, S. I. Note also travelling locomotive crane.

FIG. 40. Derrick, Belt Conveyors, Incinerators, etc. Dept. of Docks, City of New York. Reclamation of Rikers Island land under water. Grab buckets lift city refuse to screen over belt conveyor, which distributes along embankment.

FIG. 41. Lifting Towers and Belt Conveyors. J. B. King & Co., Richmond Borough Plaster Mills. Broken stone lifted by grab bucket to hoppers.

FIG. 42. Coal Hoist and Ash Conveyor. Metropolitan Street R.R. Co., 96th Street, East River.

FIG. 43. Direct Coal Unloader and Ash Conveyor. Interborough Rapid Transit Co., 74th Street, East River.

FIG. 44. DIAG. 44A. Bridge Crane and Cable Railway. N. Y. Edison Co., Shady Side, opp. 100th Street, N. Y. 200-foot span, equipped with 2-ton bucket; covers 50,000-ton storage soft coal.

FIG. 45. DIAG. 45A. Coal Storage for Power House. Kings Co. Elec. Light & Power Co., Brooklyn, N. Y. Coal hoisted 125 feet, cracked to size for automatic stoker, and deposited in storage bins over boilers.

Fig. 46. Direct Coal Unloader and Ash Conveyor, Interborough Rapid Transit Co., N. Y. Stationary booms, trolley and grab bucket. Discharges 5,000 cubic yards ashes per month. Unloads stoves and feeds 900 tons coal daily. Foot 74th Street, East River.

Fig. 47. Direct Unloader. Manhattan Elevated R.R., 129th Street and Harlem River. Moving tower, with boom and grab bucket.

Fig. 48. Automatic Self-Discharging Coke Barge. Consolidated Gas Co.
Diag. 49A.

Fig. 49. Penna. R.R. Power Station (L. I. R.R.), Long Island City. Capacity, 700 tons
Diag. 50A. per day. Storage capacity, 6,000 tons. Two-ton grab bucket lifts to hopper, delivers to crusher, to weigh hopper, to 2-ton cable cars, to coal bunkers over boilers, chutes to boilers. Ashes via chutes to ash cars, to hopper, dumping to bucket elevator to ash cable cars via chutes to ash pocket to railroad cars.

FIG. 50. Power House, L. I. R.R., L. I. City.
DIAG. 50A.

Fig. 51. Edison Electric Co., Shady Side. Direct unloader for coal storage.
Diag. 51A.

FIG. 52. Coal Tipple, D., L. & W. R.R. Piers, Hoboken. Entire coal car with load run on elevator, lifted to height of discharge chute, tipped and discharged into apron chute, discharging into boats.

FIG. 53. Coal Tipple. D., L. & W. R.R., Hoboken, N. J. Handles 2,000,000 tons per year. Elevates and dumps entire coal car.

FIG. 54. Coal Pockets, U. S. Navy Yard, Brooklyn, N. Y.

FIG. 55. Direct Unloader. Glucose Works, Weehawken. Automatic grab bucket, on adjustable boom, dumps directly into cars through hopper.

Fig. 56. Coal Tipple. B. & O. R.R., St. George, S. I.

Fig. 57. Direct Coal Unloader. J. T. Story Co., Wallabout Canal. Automatic grab bucket on stationary boom and trolley.

Fig. 58. Coal Hoister and Storage Bin. Edison Electric Illuminating Co., 65th Street, South Brooklyn. Coal hoisted from boats by grab bucket and delivered into hopper over crusher, thence by shuttle cable cars to storage. From storage by bucket conveyor to power house.

Fig. 59. Diag. 58A. Floating Coal Barge. Arbuckle Bros. Sugar Refining. Boom derrick in tower lifts coal to suspended adjustable chute, discharging in storage pile on shore.

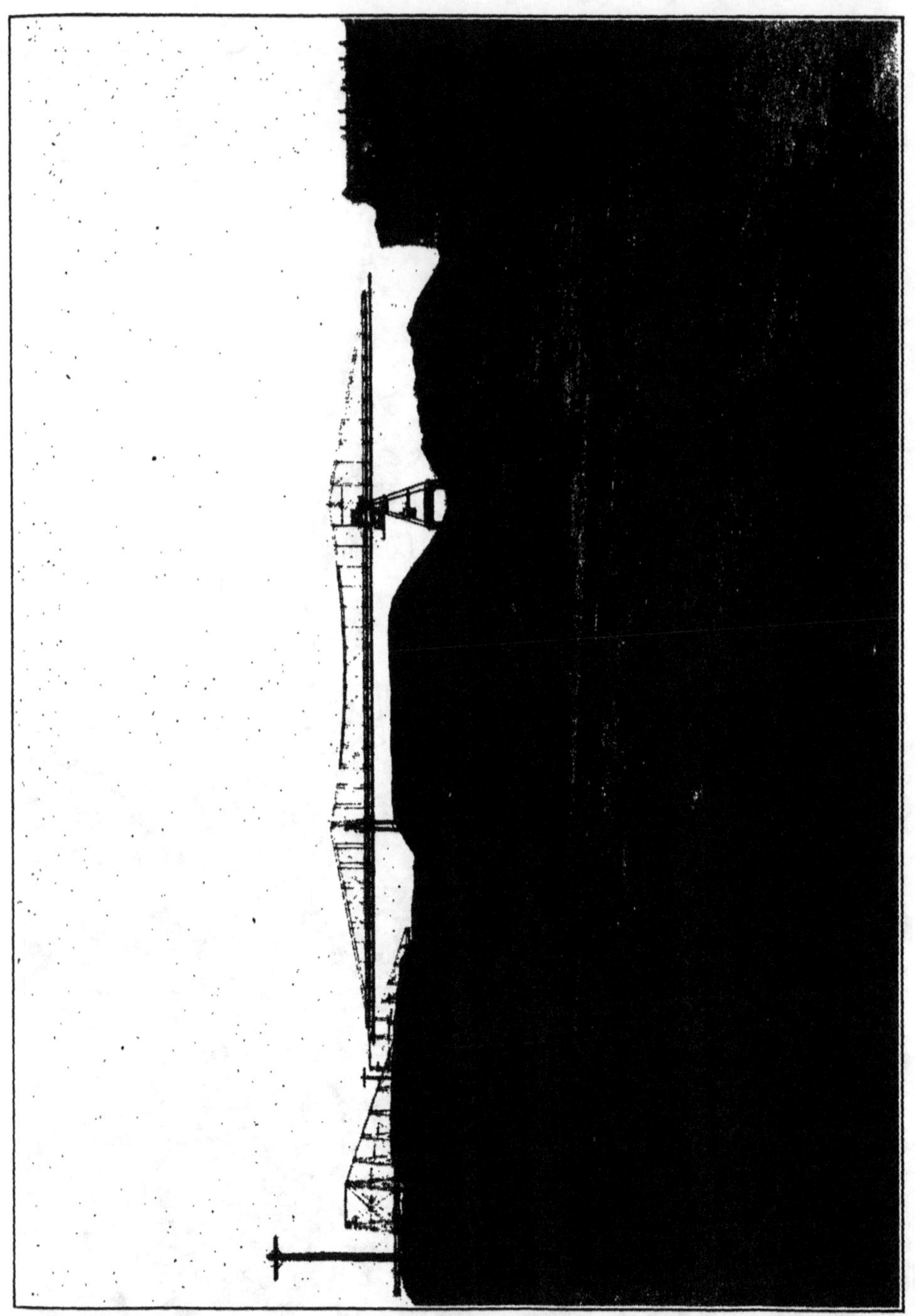

FIG. 60. Electric Bridge Tramway, with Seven-ton Automatic Grab-bucket and Man-trolley, Astoria Light, Heat & Power Company, Astoria, L. I. Capacity tons per hour, 200 to 300 tons, depending upon length of trolley travel. Length of trolley travel, 570′. Length of bridge travel, 715′. Coal taken from storage space, controlled by the gantries to space controlled by the bridge tramway, or to standard railroad cars, via hopper. Total amount of storage space controlled by the bridge tramway, 400,000 sq. ft., or 500,000 tons of coal.

Fig. 61. Electric Gantry, with Two-ton Automatic Grab-bucket. Astoria Light, Heat & Power Company, Astoria, L. I. Average capacity, 125 tons per hour. Length of boom, over-all, 192′ 6″. Coal taken from scows and delivered either to standard railroad cars, via 50-ton hoppers, or to storage space, via inboard leg of boom. Two such gantries in operation controlling a length of 715′ along the bulkhead.

Fig. 62. Water Side Station No. 1 (on left) and No. 2 (on right). N. Y.
Diags. 61A and 61B. Edison Co., foot 38th Street, East River.

Water Side Station No. 1, of the N. Y. Edison Co. Coal capacity, 100 tons per hour. One-ton bucket, hoisted by a 50 h.p. Stearns engine, delivers to a 30-ton hopper; delivers to a 24-inch crusher driven by a 20 h.p. motor; delivers to weighing hopper; delivers via chute to a 30-inch inclined belt conveyor 35 feet long, driven by a 30 h.p. motor; delivers into a coal hopper; delivers via 2-way chute into two 24-inch bucket coal conveyors running side by side up about 30′, then on an incline to top of coal bunkers and along top of coal bunkers, delivering coal into same. From coal bunkers, which have a capacity of 10,000 tons, coal is delivered to boilers via chutes. Ashes deliver via chutes with 1¼-yard ash-cars pushed by hand to 4 yd. ash hopper, delivers via gates to 24-inch belt conveyor, 115 feet long, driven by 15 h.p. motor; delivers via chute to bucket elevator 100-ft. lift, driven by a 30 h.p. motor; delivers via 2-way chute to ash pocket, 900 cu. yd. capacity; delivers via 6 chutes to scows.

Water Side Station No. 2, of the N. Y. Edison Co. Coal capacity, 100 tons per hour. One and one-half ton buckets, hoisted by a 50 h.p. engine, delivers to a 30-ton hopper; delivers to a 24-inch crusher, driven by a 20 h.p. motor; delivers to a weighing hopper; delivers via gate to 3-ton side-dumping coal cable cars drawn over the coal bunkers. There are two coaling towers, 239′ 6″ high, with hoisting boom 196 feet above deck; and two coal bunkers with a total capacity of 17,000 tons. The coal is delivered from the bunkers to boilers with chutes. Ashes deliver via chutes to 1¼-yd. ash cars drawn by a storage battery locomotive, to two 4-yd. ash hoppers; deliver via gates to two 1½-yard skips, pulled up an incline to a heighth of 115 feet above deck by two 30 h.p. motors; deliver via 2-way chute to ash-pocket, 1,200 cu. yd. capacity; deliver via 7 chutes to scows.

FIG. 63. DIAG. 62A. Cantilever Bridge with Bucket Conveyor on Dock. Interborough Rapid Transit Co., 59th Street, North River. Two conveyors are installed for handling ashes from power house to boats. Lower line of conveyors travels through tunnel under Twelfth Avenue from power house 200′ distant. Chutes at end of bridge can be moved to trim barges.

FIG. 64. DIAG. 62A. Direct Coal Unloader, Crusher and Conveyor, Interborough Rapid Transit Co., 59th Street, North River. Movable boom, trolley, and grab-bucket. Unloads and stores 900 tons of coal daily.

Fig. 65. Movable Hoisting Tower, Brooklyn Heights R.R. Co., Third Avenue Power House. 1½-ton grab-bucket; coal passes from receiving hopper through crushers, thence to pivoted bucket carrier for delivery to boiler-house bins.

Locomotive Crane lifts from barges to hopper, thence via conveyor up incline to storage pile.

Fig. 66. Direct Coal Unloader and Ash Conveyor. Metropolitan Street R.R. Co., 202nd Street, Harlem River.

Fig. 67. Typical Railroad Coal Trestle. B. & O. R.R., St. George, S. I.
Diag. 67A.

FIG. 67A. Typical Railroad Coal Pier. Railroad cars brought to top of superstructure by inclined approach; locomotive traction. Cars discharged via hoppers and chutes directly into barges or scows. Note floating coal-elevator taking cargo of coal. C. R. R. of N. J., Communipaw.

FIG. 68. N. Y. C. & H. R. R.R. Co. Elevator at 59th Street, North River.
DIAG. 70A.

FIG. 69. Grain Elevators. West Shore R.R., Weehawken, N. J., opposite 100th Street, N. Y.
DIAG. 70A.

FIG. 70. DIAG. 71A. Floating Grain Elevator, 31st Street Pier, South Brooklyn. Loads grain directly from barge to hold of steamer.

FIG. 71. Grain Elevator. Showing marine leg conveyor. Heckers' Flour Mill, Corlears Hook, East River.

FIG. 72. Grain Elevators. N. Y. Dock Co., Brooklyn.

FIG. 73. Coke Loader, Astoria Light & Power Co., Astoria, L. I. Electric belt conveyor. Car dumps into hopper under track.

Fig. 74. Diag. 75A. Barrel Conveyor and Elevator. Brooklyn Cooperage Works, Greenpoint. Capacity, 8,000 barrels per day.

Fig. 75. Barrel Hoist, J. B. Kings & Co.'s Plaster Mills.

Fig. 76. Barrie Elevator and Chute. L. V. R. R. Flour piers.

Fig. 77. Bag Chute. N. Y. Dock Co., Brooklyn.

Fig. 78. Baggage Escalator. Chelsea Piers, North River.

Fig. 79. Cargo Chute. Chelsea Piers, N. Y. City.

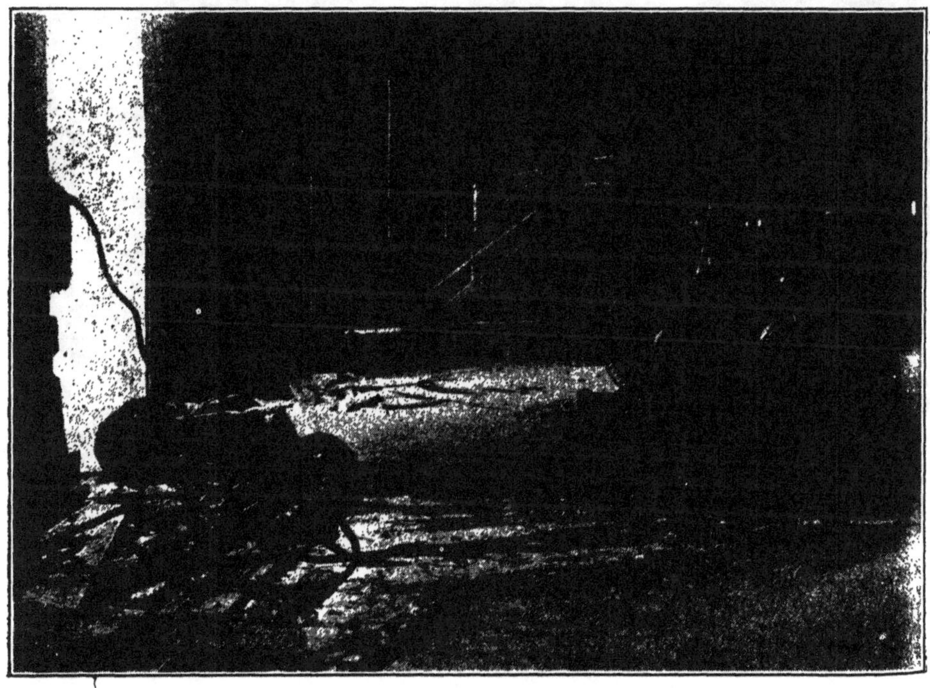

Fig. 80. Electric Cargo Hoist Winch. Fabre Line, 31st Street Pier, South Brooklyn.

FIG. 81. Tiering Machine, American Dock & Trust Co., Staten Island. Tiers general merchandise on docks and in warehouses. Electric motor.

FIG. 82. DIAG. 77A. Bag Chute, Arbuckle's Coffee Mills, East River, Brooklyn. Discharges bags from upper stories directly into cars or boats. Can be raised or lowered by tackle.

Fig. 83. Steam Derrick. Capacity, 100 tons. Department of Docks and Ferries, City of New York.

Fig. 84. Steam Derrick. North River Derrick Co., Hoboken. Capacity, 60 tons.

Fig. 85. Typical Harbor Derricks. For general lifting, wrecking, etc. Merritt, Chapman Wrecking Co., N. Y. This type of derrick varies from 10 to 200 tons capacity.

Fig. 86. Derrick Raising Sunken Tug Boat, North River.

TYPES OF STEAMSHIPS AND HARBOR CRAFT IN NEW YORK HARBOR

LIST OF HARBOR CRAFT USED FOR FREIGHT TRANSPORTATION IN NEW YORK HARBOR.

Tow boats			537
Steam derricks and hoists			123
Lighters	Steam	169	
	Steam derricks	62	
	Open	195	
	Covered	133	
	Hand derricks	87	646
Barges	Deck	566	
	Covered	385	
	Coal	362	1313
Car floats			217
Grain barges			88
Scows	Deck	683	
	Derrick	73	
	Coal	97	
	Dump	112	975
Canal boats and miscellaneous			175

FIG. 87. S.S. "Olympic" in North River. Typical transatlantic passenger steamship.

FIG. 88. S.S. "Momus," Metropolitan Line. Typical coastwise steamship.

Fig. 89. Typical Tramp Steamship.

Fig. 90. Typical Sound Steamer. S.S. "Priscilla," Fall River Line.

Fig. 91. Passenger and Freight Car Transfer Steamer "Express."
N. Y., N. H. & H. R.R. Co.

Fig. 92. Typical Railroad Car Transfer Float.

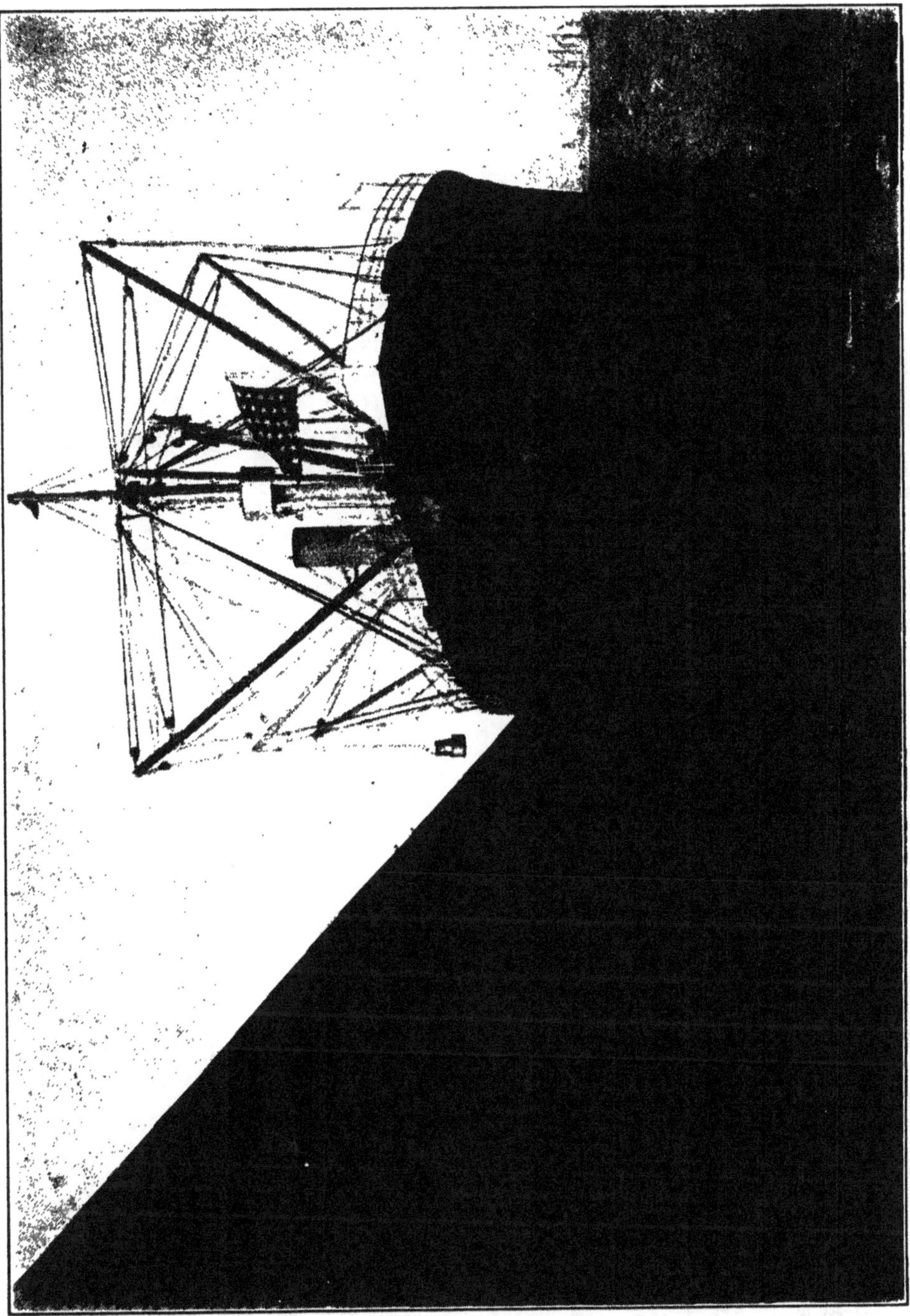

FIG. 93. S.S. "Texan," Hawaiian-American Line. Note the numerous derrick booms attached to the masts. These vessels are equipped with from twelve to twenty hoisting winches on the upper deck.

FIG. 94. Typical Ocean-going Tow Boat.

FIG. 95. Typical Steam Lighter and Tow Boat.

FIG. 96. Typical Steam Lighter, New York Harbor.

FIG. 97. Typical Deck Scow with Hand-power Derrick, New York Harbor.

Fig. 98. Types of Tow Boats, Lighters, Scows, etc., New York Harbor.

Fig. 99. Types of Covered Barges in General Use, New York Harbor.

DIAGRAMS

Showing Method of Operation of Some of the Mechanical Devices Illustrated in Previous Pages

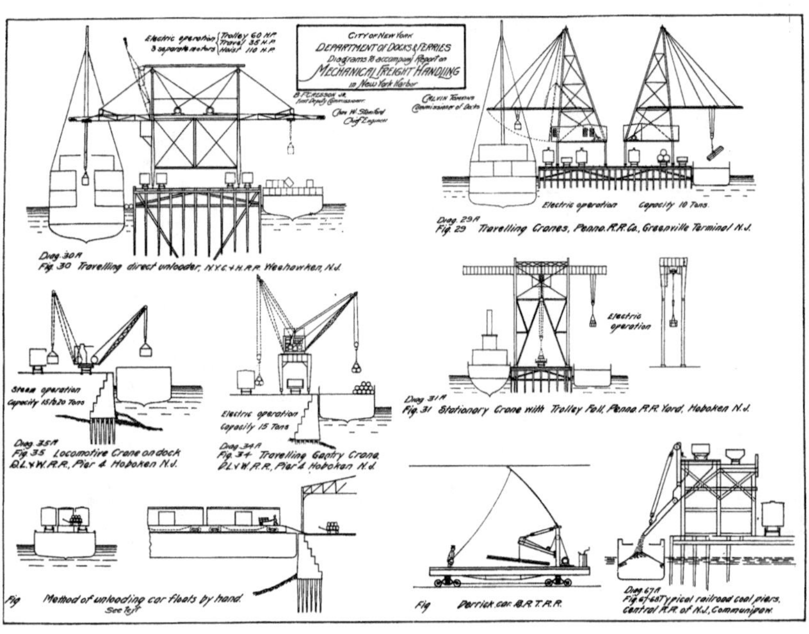

CITY OF NEW YORK
DEPARTMENT OF DOCKS & FERRIES
Diagrams to accompany Report on
MECHANICAL FREIGHT HANDLING
in New York Harbor
B.F. CRESSON JR.
First Deputy Commissioner
Chas W. Staniford
Chief Engineer
CALVIN TOMKINS
Commissioner of Docks
Electric operation
3 separate motors
Trolley 60 H.P.
Travel 35 H.P.
Hoist 110 H.P.
Diag. 30A
Fig. 30 Travelling direct unloader, N.Y.C.+H.R.R. Weehawken, N.J.
Electric operation
Capacity 10 Tons.
Diag. 29A
Fig. 29 Travelling Cranes, Penno. R.R. Co., Greenville Terminal N.J.
Electric operation
Diag. 31A
Fig. 31 Stationary Crane with Trolley Fall, Penno. R.R. Yard, Hoboken N.J.
Steam operation
Capacity 15 to 20 Tons
Diag. 35A
Fig. 35 Locomotire Crane on dock
D.L.+W.R.R., Pier 4 Hoboken N.J.
Electric operation
Capacity 15 Tons
Diag. 34A
Fig. 34 Travelling Gantry Crane
D.L.+W.R.R., Pier 4 Hoboken N.J.
Fig Method of unloading car floats by hand.
See Text
Fig Derrick car B.R.T.R.R.
Diag. 67A
Fig. 67-68 Typical railroad coal piers,
Central R.R. of N.J., Communipaw.

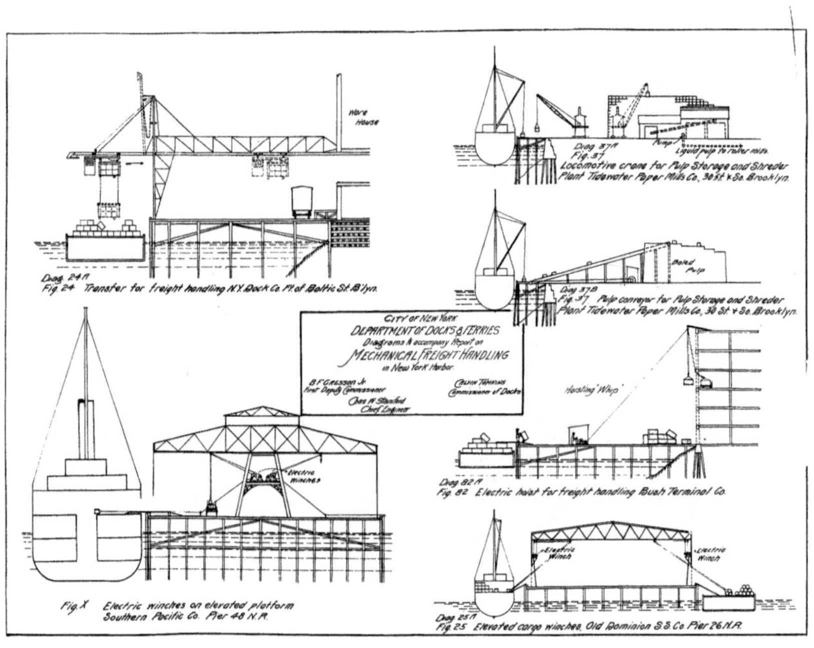
Ware House
Diag. 24 A
Fig. 24 Transfer for freight handling N.Y. Dock Co. Ft. of Baltic St. B'lyn.
Pump
Liquid pulp to roller mills.
Diag. 37 A
Fig. 37
Locomotive crane for Pulp Storage and Shreder Plant Tidewater Paper Mills Co., 30 St. & So. Brooklyn.
Baled Pulp
Diag. 37 B
Fig. 37 Pulp conveyor for Pulp Storage and Shreder Plant Tidewater Paper Mills Co., 30 St. & So. Brooklyn.
CITY OF NEW YORK
DEPARTMENT OF DOCKS & FERRIES
Diagrams to accompany Report on
MECHANICAL FREIGHT HANDLING
in New York Harbor.
B. F. Cresson Jr.
First Deputy Commissioner
Calvin Tomkins
Commissioner of Docks
Chas. W. Staniford
Chief Engineer
Hoisting "Whip"
Electric Winches
Diag. 82 A
Fig. 82 Electric hoist for freight handling Bush Terminal Co.
Fig. X Electric winches on elevated platform Southern Pacific Co. Pier 48 N.R.
Electric Winch
Electric Winch
Diag. 25 A
Fig. 25 Elevated cargo winches, Old Dominion S.S. Co. Pier 26 N.R.

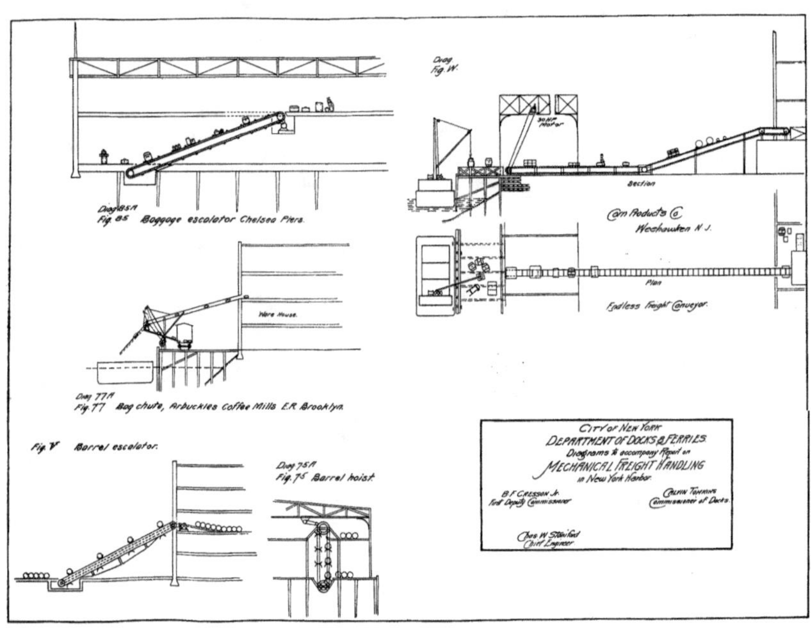
Diag 85A
Fig. 85 Baggage escalator Chelsea Piers.
Diag.
Fig. W.
30 HP
Motor
Section
Corn Products Co
Weehawken N.J.
Plan
Endless Freight Conveyor.
Ware House.
Diag 77A
Fig. 77 Bag chute, Arbuckles Coffee Mills E.R. Brooklyn.
Fig. V Barrel escalator.
Diag 75A
Fig. 75 Barrel hoist.
CITY OF NEW YORK
DEPARTMENT OF DOCKS & FERRIES.
Diagrams to accompany Report on
MECHANICAL FREIGHT HANDLING
in New York Harbor.
B.F. CRESSON Jr.
First Deputy Commissioner
CALVIN TOMKINS
Commissioner of Docks.
Chas. W. Staniford
Chief Engineer

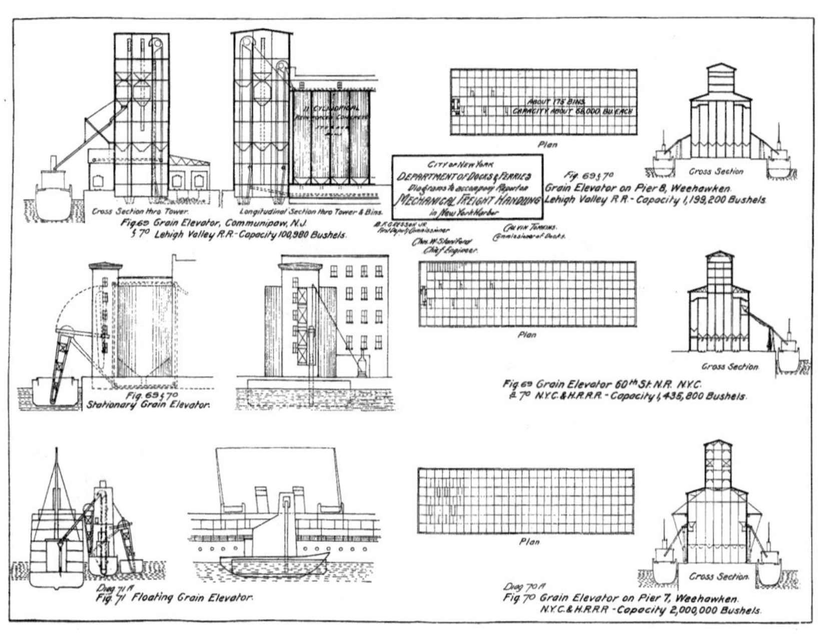
11 CYLINDRICAL
REINFORCED CONCRETE
Cross Section thro Tower.
Longitudinal Section thro Tower & Bins.
Fig. 69 Grain Elevator, Communipaw, N.J.
& 70 Lehigh Valley R.R. - Capacity 100,980 Bushels.
ABOUT 175 BINS
CAPACITY ABOUT 68,000 BU. EACH
Plan
Cross Section
CITY OF NEW YORK
DEPARTMENT OF DOCKS & FERRIES
Diagrams to accompany Report on
MECHANICAL FREIGHT HANDLING
in New York Harbor
B. F. CRESSON JR.
First Deputy Commissioner
Chas. W. Staniford
Chief Engineer.
CALVIN TOMKINS.
Commissioner of Docks.
Fig. 69 & 70
Grain Elevator on Pier 8, Weehawken.
Lehigh Valley R.R. - Capacity 1,199,200 Bushels.
Fig. 69 & 70
Stationary Grain Elevator.
Plan
Cross Section
Fig 69 Grain Elevator 60th St. N.R. N.Y.C.
& 70 N.Y.C. & H.R.R.R. - Capacity 1,435,800 Bushels.
Plan
Cross Section
Diag 71 A
Fig. 71 Floating Grain Elevator.
Diag 70 A
Fig 70 Grain Elevator on Pier 7, Weehawken.
N.Y.C. & H.R.R.R. - Capacity 2,000,000 Bushels.

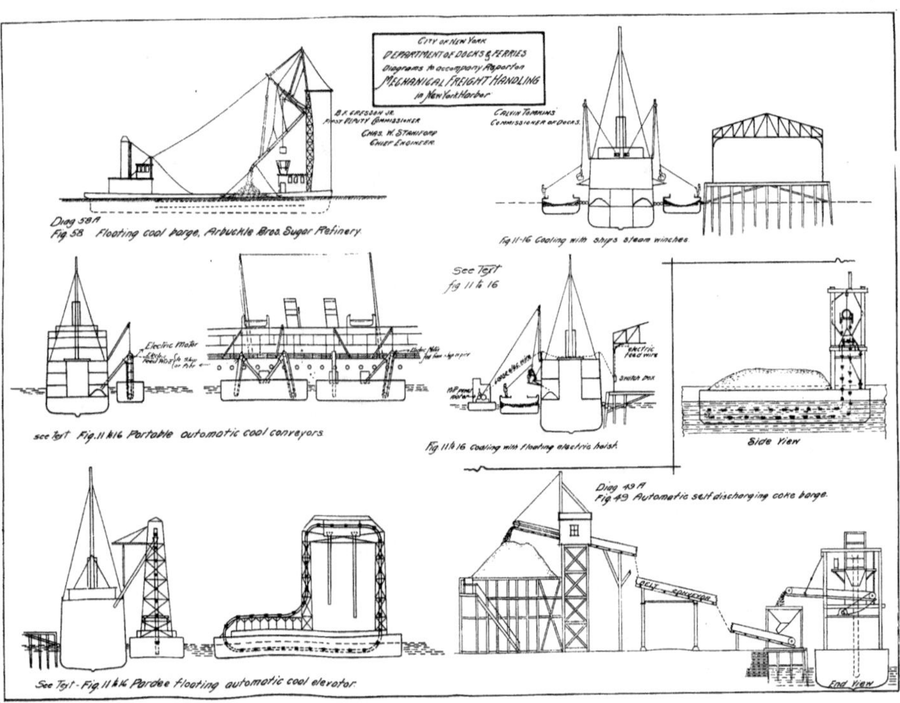
CITY OF NEW YORK
DEPARTMENT OF DOCKS & FERRIES
Diagrams to accompany Report on
MECHANICAL FREIGHT HANDLING
in New York Harbor
B. F. CRESSON JR.
FIRST DEPUTY COMMISSIONER
CHAS. W. STANIFORD
CHIEF ENGINEER
CALVIN TOMKINS
COMMISSIONER OF DOCKS.
Diag 58 A
Fig. 58 Floating coal barge, Arbuckle Bros. Sugar Refinery.
Fig. 11-16 Coaling with ships steam winches.
Electric Motor
See Text
fig 11 to 16
electric feed wire
Switch box
Side View
see Text Fig. 11 to 16 Portable automatic coal conveyors.
Fig. 11 to 16 Coaling with floating electric hoist.
Diag 49 A
Fig. 49 Automatic self discharging coke barge.
BELT CONVEYOR
End View
See Text - Fig. 11 to 16 Pardee floating automatic coal elevator.

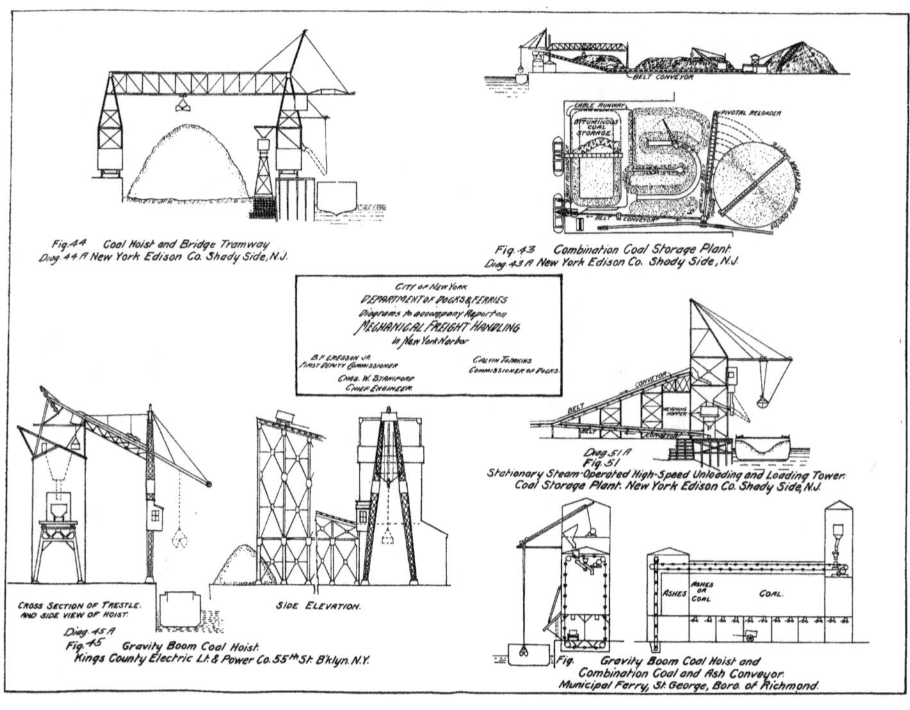

Fig. 44 Coal Hoist and Bridge Tramway
Diag. 44 A New York Edison Co. Shady Side, N.J.

Fig. 43 Combination Coal Storage Plant.
Diag. 43 A New York Edison Co. Shady Side, N.J.

Diag. 51 A
Fig. 51
Stationary Steam-Operated High-Speed Unloading and Loading Tower.
Coal Storage Plant. New York Edison Co. Shady Side, N.J.

Diag. 45 A
Fig. 45 Gravity Boom Coal Hoist.
Kings County Electric Lt. & Power Co. 55th St. B'klyn. N.Y.

Fig. Gravity Boom Coal Hoist and
Combination Coal and Ash Conveyor.
Municipal Ferry, St. George, Boro. of Richmond.

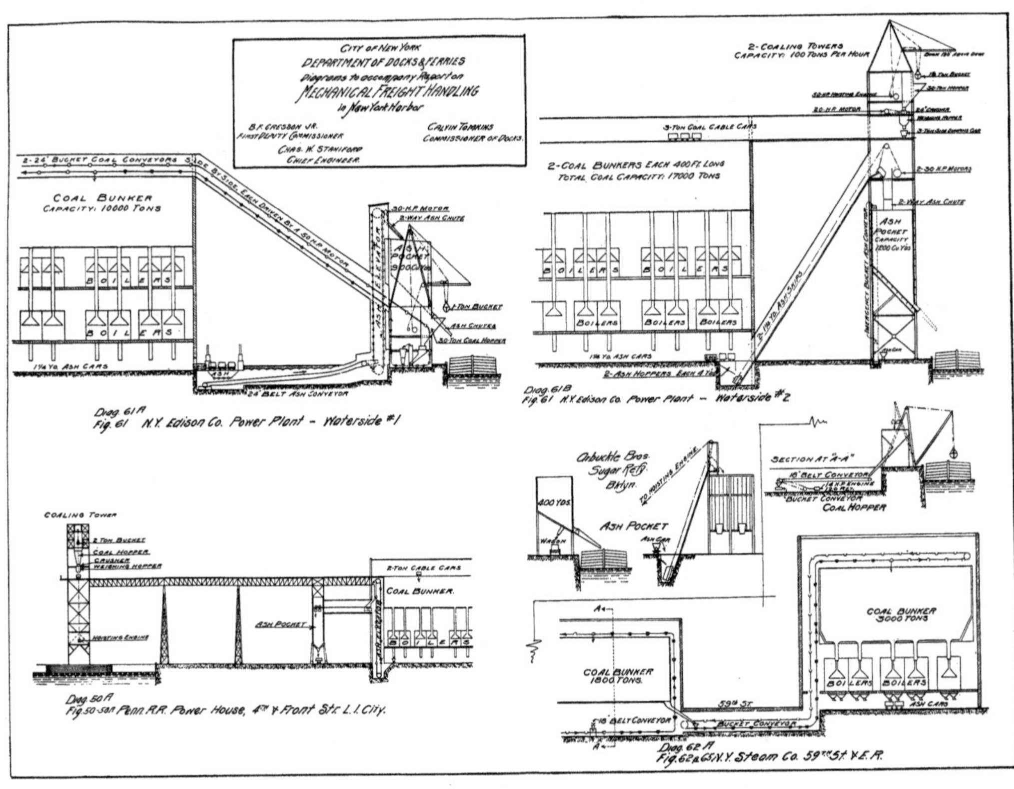

CITY OF NEW YORK
DEPARTMENT OF DOCKS & FERRIES
Diagrams to accompany Report on
MECHANICAL FREIGHT HANDLING
in New York Harbor
B. F. CRESSON JR.
FIRST DEPUTY COMMISSIONER
CALVIN TOMKINS
COMMISSIONER OF DOCKS.
CHAS. W. STANIFORD
CHIEF ENGINEER.
2-24" BUCKET COAL CONVEYORS SIDE BY SIDE, EACH DRIVEN BY A 50 H.P. MOTOR
COAL BUNKER
CAPACITY: 10000 TONS
BOILERS
BOILERS
1¼ YD. ASH CARS
ASH
24" BELT ASH CONVEYOR
ASH ELEVATOR
30-H.P. MOTOR
2-WAY ASH CHUTE
ASH POCKET 900 Cu. Yds
1-TON BUCKET
ASH CHUTES
30-TON COAL HOPPER
Diag. 61 A
Fig. 61 N.Y. Edison Co. Power Plant - Waterside #1
2-COALING TOWERS
CAPACITY: 100 TONS PER HOUR
50-H.P. HOISTING ENGINE
30-TON HOPPER
20-H.P. MOTOR
24" CRUSHER
WEIGHING HOPPER
3-TON COAL CABLE CARS
2-COAL BUNKERS EACH 400 FT. LONG
TOTAL COAL CAPACITY: 17000 TONS
2-30 H.P. MOTORS
2-WAY ASH CHUTE
ASH POCKET CAPACITY 1200 Cu. Yds
EMERGENCY BUCKET ASH CONVEYOR
2-1¼ YD. ASH SKIPS
BOILERS
BOILERS
BOILERS
1¼ YD. ASH CARS
ASH CAR
2-ASH HOPPERS EACH 4 YDS
Diag. 61 B
Fig. 61 N.Y. Edison Co. Power Plant -- Waterside #2
Arbuckle Bros. Sugar Refy. Bklyn.
400 YDS.
WAGON
ASH POCKET
TO HOISTING ENGINE
ASH CAR
SECTION AT "A-A"
18" BELT CONVEYOR
14 H.P. ENGINE
BUCKET CONVEYOR
COAL HOPPER
COALING TOWER
2 TON BUCKET
COAL HOPPER
CRUSHER
WEIGHING HOPPER
HOISTING ENGINE
ASH POCKET
2-TON CABLE CARS
COAL BUNKER.
BOILERS
Diag. 50 A
Fig. 50-50A Penn. R.R. Power House, 4th & Front Sts. L. I. City.
COAL BUNKER
3000 TONS
COAL BUNKER
1800 TONS.
59th ST.
18" BELT CONVEYOR
BUCKET CONVEYOR
BOILERS
BOILERS
ASH CARS
A
A
Diag. 62 A
Fig. 62 & 63 N.Y. Steam Co. 59th St. & E.R.

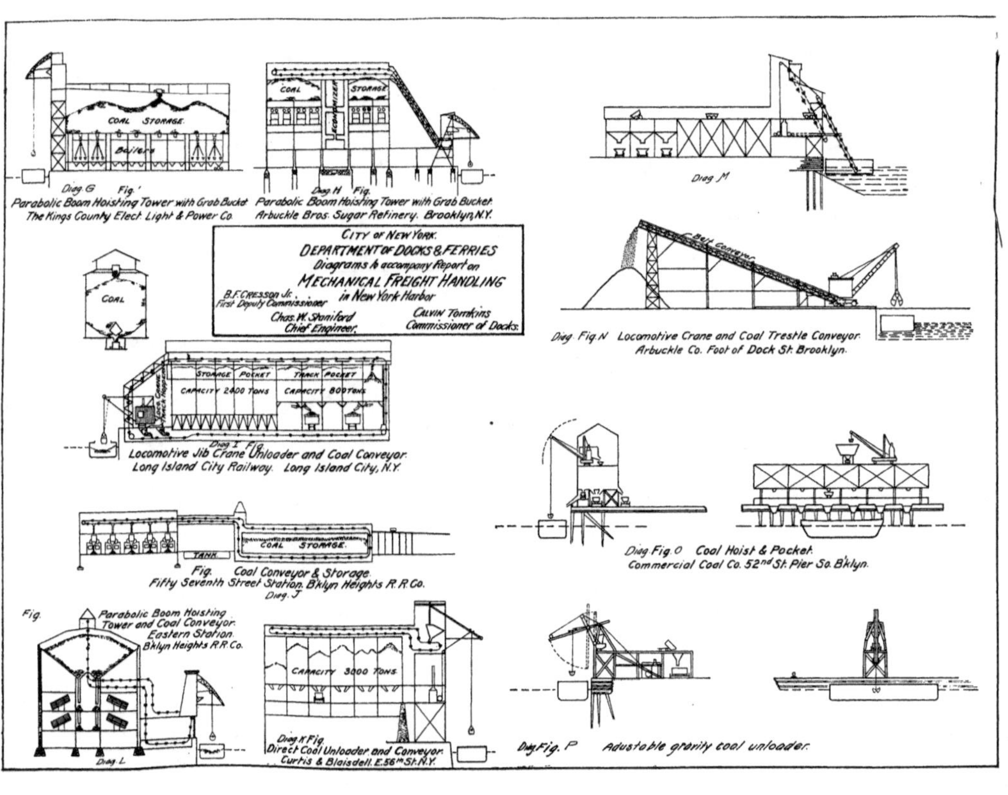

COAL STORAGE
Boilers
Diag. G Fig.
Parabolic Boom Hoisting Tower with Grab Bucket
The Kings County Elect. Light & Power Co.
COAL
ECONOMIZER
STORAGE
Diag. H Fig.
Parabolic Boom Hoisting Tower with Grab Bucket
Arbuckle Bros. Sugar Refinery. Brooklyn, N.Y.
Diag. M
CITY OF NEW YORK.
DEPARTMENT OF DOCKS & FERRIES
Diagrams to accompany Report on
MECHANICAL FREIGHT HANDLING
in New York Harbor
B.F. CRESSON Jr.
First Deputy Commissioner
Chas. W. Staniford
Chief Engineer.
CALVIN TOMKINS
Commissioner of Docks.
COAL
Belt Conveyor
Diag. Fig. N Locomotive Crane and Coal Trestle Conveyor.
Arbuckle Co. Foot of Dock St. Brooklyn.
STORAGE POCKET
TRACK POCKET
CAPACITY 2400 TONS
CAPACITY 800 TONS
Diag. I Fig.
Locomotive Jib Crane Unloader and Coal Conveyor.
Long Island City Railway. Long Island City, N.Y.
Diag. Fig. O Coal Hoist & Pocket.
Commercial Coal Co. 52nd St. Pier So. B'klyn.
TANK
COAL STORAGE
Fig. Coal Conveyor & Storage.
Fifty Seventh Street Station. B'klyn Heights R.R. Co.
Diag. J
Fig.
Parabolic Boom Hoisting
Tower and Coal Conveyor.
Eastern Station.
B'klyn Heights R.R. Co.
Diag. L
CAPACITY 3000 TONS
Diag. K Fig.
Direct Coal Unloader and Conveyor.
Curtis & Blaisdell. E. 56th St. N.Y.
Diag. Fig. P Adustable gravity coal unloader.

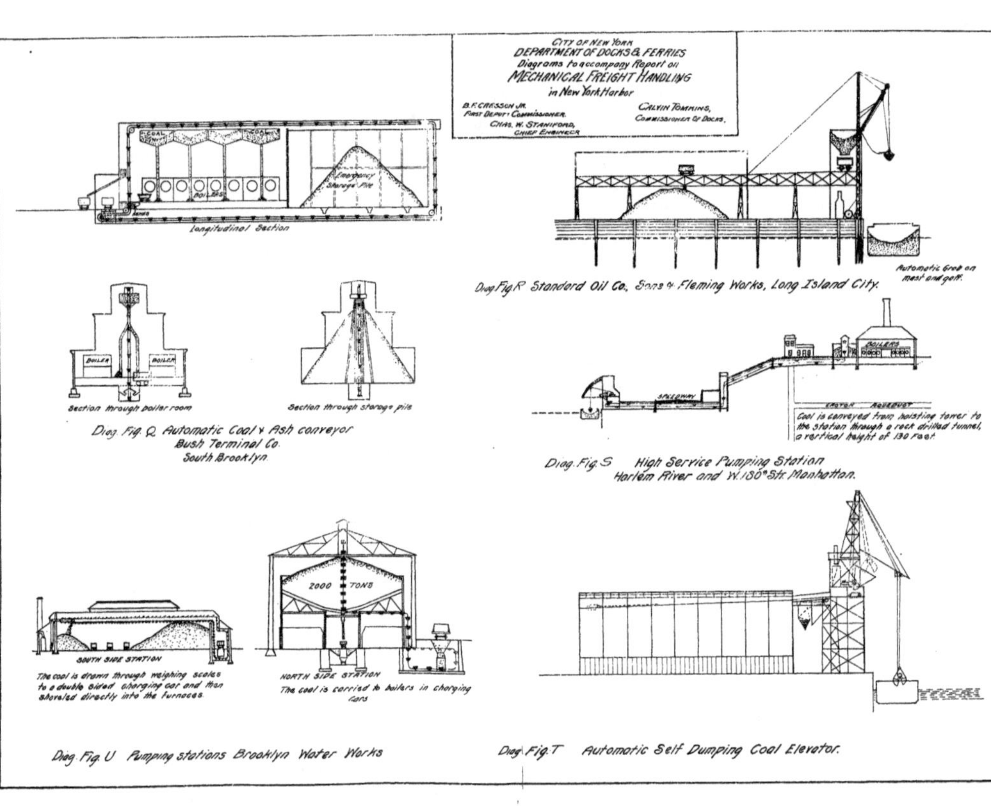
CITY OF NEW YORK
DEPARTMENT OF DOCKS & FERRIES
Diagrams to accompany Report on
MECHANICAL FREIGHT HANDLING
in New York Harbor
B. F. CRESSON JR.
FIRST DEPUTY COMMISSIONER.
CHAS. W. STANIFORD,
CHIEF ENGINEER
CALVIN TOMKINS,
COMMISSIONER OF DOCKS.
COAL
COAL
BOILERS
Emergency Storage Pile
Longitudinal Section
BOILER
BOILER
Section through boiler room
Section through storage pile
Diag. Fig. Q. Automatic Coal & Ash conveyor
Bush Terminal Co.
South Brooklyn.
Automatic Grab on mast and gaff.
Diag. Fig. R Standard Oil Co., Sons & Fleming Works, Long Island City.
BOILERS
SPEEDWAY
Coal is conveyed from hoisting tower to the station through a rock drilled tunnel, a vertical height of 130 feet.
Diag. Fig. S High Service Pumping Station
Harlem River and W. 180th St. Manhattan.
2000 TONS
SOUTH SIDE STATION
The coal is drawn through weighing scales to a double sided charging car and then shoveled directly into the furnaces.
NORTH SIDE STATION
The coal is carried to boilers in charging cars
Diag. Fig. U Pumping stations Brooklyn Water Works
Diag. Fig. T Automatic Self Dumping Coal Elevator.

www.ingramcontent.com/pod-product-compliance
Lightning Source LLC
LaVergne TN
LVHW010613110826
845149LV00003B/892

* 9 7 8 1 4 1 8 1 8 7 1 2 5 *